I0408982

CONTENTS

MARIKA SOPHIS

As Above So Below

Bridging the Gap Between Science & Spirituality

Marika Sophis

MARIKA SOPHIS

Taking God Out of the Equation is not a pre-requisite to embracing Science.

Inspired by I.C. Elohim

About The Author

Marika Sophis is certified in Solar Spectrum Sound Therapy in the modalities of Tuning Fork therapy, Chromatherapy, Sympathetic Vibratory Resonance & Brainwave entrainment. As an experienced practitioner of Metaphysics, she engages in self healing techniques utilizing the modalities of Sound, Sympathetic Resonance & Bio-field Energy through a comprehension of the Scientific Processes of Nature. Marika Sophis is a seeker of Truth, Wisdom & Light. She is a natural philosopher and an active researcher of many years. Also the published author of Awaken & Ascend & As Above So Below, Macrocosmic Reflections in a Microcosmic Mirror.

Dedication

I dedicate this book to the aspiration of a non judgmental, intelligent, logical exploration of alternative scientific theories. This work is dedicated to the many quintessential correlations between the Scientific Processes of Nature and Universal Law. A Spiritual Revolution is unfolding in our modern age. Marvelously amazing new evidence and discoveries being brought to our attention by the James Webb Space Telescope may help solve the mysteries of the Universe. This highly advanced technology is aptly equipped for the task. The brilliant minds of the academic scientific community are key players in the dawning age of light. Star gazers everywhere examine images returned to earth in wonder and astonishment.

What new insights about our radiant Cosmos will be revealed to humanity? Can a true comprehension of our luminescent Universe define our purpose and existence? Yes, it most certainly can. Such is the wondrous Nature of our Creator. Astrophysicists, cosmologists, cosmogonists, physicists, biologists, chemists, plasma physicists, astronomers and alchemists of the world, you are next in line to become the modern heroes of the human race. The Aquarian Age is an age of logic, knowledge, reason and Light. This pivotal juncture in linear time was indeed prophesied thousands of years ago. Are we ready for plasma physics and electric universe teachings to be taken more seriously?

JWST has taken center stage in the academic scientific arena. Bridging the gap between science and spirituality is inevitable. This revolution, has already initiated within the rapidly growing, expanding New Age Community. This movement is one of unity, oneness, enlightenment, love and compassion.

The Aquarian Age heralds forth divine knowledge and truth. Light comes and more light comes. Validating metaphysics, demystifying science and supplying humanity with logical, rational evidence of universal Law at work in Nature has become a spiritual obligation. The veil of secrecy naturally gives way to the radiant Light now entering the minds of spiritual people. A mass awakening is the effect of this divine cause. Supplying humanity with well researched, proven, accepted scientific facts and truths of an origin theory that corresponds to Natural Laws of the Universe, as we already know them to be, is the spark that kindles the flame. This light paves the path to brilliant radiance derived from the gift of knowing. In this way, the world will be changed for the better. In effect, the sympathetic resonance of human and planetary consciousness can be raised. Educated, unrestricted truth regarding our existence serves to unify humanity in oneness with purpose, faith and hope.

Out in the expansive Cosmos, James Webb gazes with his keen, sharp golden eyes, peering deep into the starry sky. Penetrating Space with revealing red light, illuminating that which was once blocked from sight. Into the lofty, stellar heights this Golden Boy can see. Can the ultimate question be answered? Why are we here? Who are we? How did all things come to be? I'm counting on James Webb to help us solve these mysteries. In hope of an enlightened future for all, I dedicate this book to thee.

Marika Sophis

Table Of Contents

the Radical, Expansive Perspective of Halton Arp.

Chapter 9- Objectively and Subjectively Reviewing the Observable Evidence Presented by Astronomer Halton Arp the Galileo of Palomar.

Chapter 10- Connecting the Quintessential Dots Between Arps Concept of Galaxy Formation, Metaphysical Creation Models and Universal Law.

Chapter 11- A Simple Overview of Radionics and Electromagnetic Morphic Energy Fields.

Chapter 12- Reviewing the Benefits of the Most Advanced Bio-field Meditation System Available on Todays Global Market.

Chapter 13- Mind Over Matter. Learning How to Alter Our Brainwave States and Heart Rate at Will.

Acknowledgments

In highest respects, I acknowledge the late physicist Wal Thornhill as one of the most brilliant scientific minds of this generation. With further reverence I acknowledge the Galileo of Palomar, Halton Arp. A magnificently inquisitive, star gazing explorer, who never stopped asking questions about the origin of our Universe. Much like Wal Thornhill, he sought stellar light for his entire scientific life. I further extend well earned acknowledgements to all under appreciated scientific minds, whose phenomenal abstract theories and concepts go unrecognized by mainstream science, no matter how heavily evidenced their ideas are. The radical concepts of Immanuel Velikovsky regarding planetary interactions were discarded by the professional scientific community in great haste. Yet, his teachings would have changed the course of history and the field of science as we know it.

Immanuel Velikovsky redefined our understanding of gravity. Wal Thornhill made this explosive body of evidence his life's work. Together with David Talbott, he founded the Thunderbolts Project. Thornhill spent much of his academic career inside the mind of Velikovsky, resurrecting him from within the vector field. The effect of this cause, was the revitalization of the largely unaccredited, brilliant scientific work of this magnificent, legendary mind. As such, the discoveries of Velikovsky were brought back to the attention of humanity. Once more, Velikovsky studied the stars, planets and gravity only this time, it was through the perceptive, observant eyes of Wal Thornhill. Together, these two great minds now unite as one force within the Collective Consciousness. Herein mentioned are only some of the brave soldiers of science who exceedingly attempted to drive home their valid, heavily evidenced, highly significant scientific concepts. Each of them met with tremendous resistance, rejection, judgement and lack of professional funding to further prove their research. Despite the many brick walls and slammed doors they confronted, these minds never ceased to diligently fight their causes. Let not their faithful efforts be fruitless or in vain. Their spirits were never broken. Their beliefs were never shaken. Their faith remained firm against the pressures of those who fight to maintain the known construct of accepted, standard model scientific paradigms. Thus, the modern scientific community, intelligent academics and students in Universities everywhere have yet to fully benefit from the profound knowledge and truths that they, in combination have accumulated for humanity

Bridging The Gap Between Science & Spirituality

Science is for everyone. Developing a comprehension of the scientific processes of Nature, draws one closer to the Absolute. It is imperative that humanity comes to understand that all biological creation in existence, has been produced through Universal Law. In this new age of Light, science will inevitably, finally be accredited as the work of our Creator. Loyal Christians who worry or fret over the combination of religion and science, should truly be permitted to openly explore the majestic methods of God guilt free. God and science in reality, merge as One divine, omnipresent force. The truth truly is out there. Science explains the methods utilized by our Source to manifest everything in existence that we see and cannot see all around us. Science would not even exist if not for the need to manifest and explain the great creative work of our Source. This is what all of humanity may soon come to acknowledge. In comprehending this we become capable of raising the harmonic frequency of our planet and our human race. In this way, each of us contributes to the global spiritual awakening that is already in progress. The Aquarian Age of Light is upon us. Humanity is waking up all across the globe.

The time has come to bridge the gap between Science and Spirituality. In my writings I am attempting to de-mystify otherwise complicated, relevant, scientific knowledge. Fundamentally, the necessity for modern science to simplify complicated scientific concepts, theories, paradigms, physics and cosmic processes is long overdue. If science is truly for everyone, which it ultimately is, average people should know the privilege to learning of and benefiting from quintessential scientific teachings on a personal level. Through knowledge such as this, we can learn to incorporate essential academic science, metaphysical science, mystical science and sacred wisdom into our daily lives. This action causes the effect of enlightenment. In the light of knowledge, humanity has an opportunity come to an understanding of why we are here, how we came to be here and what our purpose of existence is.

Forward

Stepping Stones. A New Perspective By Which To Establish Spiritual Order Out Of Cosmic Chaos

Who is our Creator? What is the Source of Consciousness? The Absolute and the force of consciousness can be experienced in our biological world. The Primordial Consciousness that generated space itself along with all scientific processes in the cosmos can be traced all the way back to the Cosmic Egg and the initial construction of the Seed of Life. Once God becomes the fundamental archetype of the perfect sphere with a dot in the middle, we have reached the dawn of Creation. Humanity cannot see beyond this, in the prior, as we are part of what came after. This is the origin point of primordial creation in the area of the Vector Field where all we see and know of exists. Inside our physical bodies, our Soul Self is capable of communing with the spirit forces of Nature. We cannot see into the realm of the Absolute nor can anyone come back to tell the story once their Soul returns home to the abode of God-Mind. To accept this detail is to know on a deep personal level that the closest we can come to our Source is to develop a relationship with our divine, immortal Inner Self. This spiritual aspect of our human constitution is eternally connected to our Source.

In an electric universe, smaller cluster galaxies are born as ejections from a plasmoid. A plasmoid is a luminous body scientifically referred to as an active galactic nuclei. The ejected quasi stellar object is known as a quasar. In the findings of Halton Arp, quasars are born as high red shift cosmic anomalies. In the electric universe theory, high red shift equates to youth. In great contrast, the standard scientific model, redshift is an indicator of great age and vast distance. Highly advanced technology now exists that is capable of aiding scientists in facilitating a better understanding of the Universe we live in. There is so much more to our story than what we already know as a race. In addition, much of what we know about Science may have sadly been misinterpreted. All space is filled with life and activity. All creation is bound by Universal Law that is regulated and governed by the Spirit Forces of Nature. All is of sacred, intelligently organized, magnificent design. Everything in Nature historically repeats itself. All is modeled through the primordial patterns of Creation. Our Earth is a spec of space dust in a massive galaxy filled with dust and planetary objects. All creation is patterned through fractals, sacred geometric archetypes, harmonic frequencies, the Fibonacci sequence, the golden ratio and so on. These omnipresent systems of manifestation have constructed the divine architecture of the Holographic Universe That we live in. Bearing in mind that all aspects of creation are powered by scientific processes in an infinite state of repetition do you think it possible for our human race to be the only human race in existence? Are we alone in the cosmos? From the vantage point of Universal Law is this even a possibility? In reality, the very notion itself conclusively defies the laws of science. It further defies the ancient metaphysical and alchemical truths that we already do know regarding Nature, Creation and Universal Law. In short, we can't possibly alone. Why not? An simple understanding of metaphysical science shows us why this is impossible. Essentially, that's not how the creative, reflective mirror of Nature works. Everything in Nature repeats itself.

Introduction

In March 2023, I wrote my 1st book *Awaken and Ascend.* In April/May, I wrote my 2nd book, which is book 1 in this current 3 volume series entitled *As Above, So Below. Macrocosmic Reflections in a Microcosmic Mirror.* That was the fore runner to the book you are currently reading now. By November the final book of this series will be progress. These are my 1st published works. On a personal level, it was imperative to me that these publications spring to life during the Chinese Zodiac Year of the Rabbit. The profound healing energy of this sign meets current requirements of humanity. My alchemical message is only a guide based on my knowledge and experience. Hopefully it will be of spiritual service to you in some form or another. Doing the real work of attaining spiritual evolution is up to you. Living a healthy life in a higher state of vibration and expanded states of consciousness is a personal choice. I have spent 30 years of my life searching for the truth in a world filled with lies, half truths and grand misconceptions. The tools that I have tested, utilized and resources found in these literary works are rooted in good information. In regard to academic research, I have always had the ability to distinguish the truth from a lie. This gift of a divine, internal sense of knowing tends to serve me well. To demystify this talent is to comprehend that there are three energy centers in our biological earth vessel that warn of us of inharmonious vibrational feedback.

We all have three warning centers that alert us in the event of encounters with various forms of danger and unhealthy activity. Unbeknownst to most, these three speak to us through the power of frequency at all times. Fundamentally our heart, our gut and our mind or intuition constitute these key centers. Metaphysically speaking, I am referring to our solar plexus chakra, our heart chakra and our brow or third eye chakra. When all three are in agreement on any given academic subject or positive response to external stimuli, pay very close attention to the matter at hand. This is a clear sign that you are in the presence of truth in some form. Sharing my story along with the proper spiritual wisdom I have painstakingly accumulated over the years and compiled into one series is my selfless service to humanity. This information has all been dredged out of the public domain. Ultimately, it was yours to begin with. If revealing any divine truths that I have discovered in this way, aids others on their own path to light, I have done a good deed on behalf of our Source and Nature. That being said, this message may not be for everyone. If so that is perfectly fine. True wisdom essentially comes from the Inner Self. This aspect of our being is always connected to Source energy. The Spirit Forces of Nature and the unconditional love of the Absolute can be found everywhere in existence. By tapping into this frequency you can gain access to universal knowledge. The ability to access the wisdom of the collective consciousness is a birth rite of the human race. The expansive halls of records were in fact, complied through the incarnate cycles of every human who ever lived. The collective mind is composed of the memories and experiences of every age and every generation of our race. As such, the wisdom contained therein belongs to all of us. We all generated it together as a species. In addition, knowledge of metaphysics, alchemy and spiritual science belongs to every member of humanity past and present. This knowledge too, resides within the expansive consciousness of the collective mind.

Consequently, the scientific processes of Nature have been made so complicated to comprehend that the majority of people don't even care to try. De-mystifying and simplifying confusing scientific concepts and models of creation, facilitates development of an understanding of our spiritual, biological existence. Bridging the gap between science, spirituality and religion is long overdue. In truth these forces are not separate. They are quintessentially woven together as one powerful, omnipresent force governed by Universal Law. The average person may not wish to study quantum physics, biology, astronomy nor chemistry. However, each of us can still benefit greatly from acquiring a basic knowledge of universal scientific systems of manifestation. Such knowledge is relevant to enhancing and improving our life experiences, daily existence, relationships and internal connection to our Creator. All things are connected. This knowledge is key. Each member of the human race is a crystallized biological by-product of the scientific processes of Nature. We embody the unconditional love, consciousness and essence of our creator. We are naturally, eternally driven to strive towards correct ethics, justified, fair conduct and proper action. Our choices are governed by Universal Law.

Our perceived individual consciousness is part of the mind of God. The Absolute is the primordial, ultimate source of Consciousness. When our incarnate journeys are complete, our spiritual soul consciousness makes its journey back home to God-Mind. Our consciousness is pre-programmed to make this journey when our soul has spiritually developed and evolved to its fullest maximum potential. How do we evolve to this state of purity? Via the ongoing learning process of biological existence inside of our material bodies, our many earth vessels. In effect, we are divine immortal spirit beings having a mortal human experience. It is not the other way around as we have been taught. As in Biblical teachings we do return to the realm of God when we die. As in Eastern Philosophical teachings, we do reincarnate a great number of times before our soul is purified and evolved enough to make this journey home to God. Consequently both philosophies are essentially correct. God, Brahma, Krishna, Allah, YHWH, Jehovah, Source, Creator, Absolute, Amun/Atum/Ra, Zeus,

Kronos the All and so on, are many titles which describe the One Creator. All known traditional, positive ethical, moral, spiritual religious systems of belief simply lead to the same ultimate, divine, omnipresent Almighty Source. The reincarnate process was intelligently designed as a vehicle for soul and consciousness evolutionary development. Through a search for Light, the quest for knowledge and truth, spiritual devotion rooted in selfless service to fellow members of our race and unconditional love, we all find our way back home to God. Eventually. This final step comes only after a vast number of lifetimes have been experienced and all necessary lessons have been learned. In fact, it takes a multitude of lifetimes to properly prepare our soul to meet the divine Force who is responsible for everything.

Is God a man? Is God a woman? In truth, in Alchemy and metaphysics the nature of biological existence is rooted in duality. As such, our Creator would contain the essence of both male and female, in equal parts and equal quantities akin to asexual energetic aspects of manifestation. Thus, all dual forms of existence be they male or female both emerge from the One Divine Source. This information is key. Equality within our human race is a reflection of the essence of the Absolute. In all honesty, our Creator likely doesn't even care what we call it. It simply just wants us to call. Ultimately, our Source answers to a lot of names assigned to it by various cultures of the world. So be it. There is a spiritual path to the light of truth for everyone. We all have the power to achieve the most amazing results in regard to personal growth. This result is the effect of choosing to live in a healthy, higher vibrational state of mind and being. The literary works that I am currently producing at the rate of speed at which this mission is being accomplished speaks for itself. Where is all of this fluid information, energy and productivity coming from? How am I doing all of this while also working a full time job in an environment that has presented endless outside pressure for several years? What is the mystery behind this super phenomenal drive? May some of this resiliency, knowledge and stamina also be coming from my biological and spiritual alignment to Inner Self, to the earth, to the scientific spiritual forces of Nature and to a higher harmonic frequency? Absolutely.

Once we open our hearts to embrace the self within, we can literally feel the omnipresent, unconditional love of our Source all throughout us and all around us. This sensation can be sensed and experienced within our chakra energy centers, bio-field energy, consciousness, sympathetic astral bodies and our biological earth vessel. Our decision to reach this lofty level of awareness causes our spiritual imbalances, vices and inner psychological demons to rear their ugly heads and demand salvation. Why? Because negative energies and unhealthy, dark, hideous things are always disturbed , dredged out and removed by the light of Spirit. Amen. The negative energy and discomfort that we experience in life is very often rooted in the energy of resistance. This resistance often stems from various self defeating patterns of behavior. Weaknesses are rooted in negative and disempowering states of mind and being. We subconsciously create our reality at all times with every thought we have and every breath we take. If we depend on our environment, other people, life circumstances and other things outside of ourselves to make us deeply internally happy, at one with our Inner Self and our Source, this will never happen. Dream on. This force doesn't come from outside of ourselves. It comes from within. We must choose positive change of our own free will. We must have the desire to commit to self improvement. Initially this begins with a choice to become the very best version of ourselves that we can potentially strive to be at our most maximum potential. Whenever any member of the human race creates a better version of themselves, it benefits the entire biological world that we live in. It raises the harmonic frequency the human race as a whole. An excellent contribution indeed. A deep sense of joy, peace, stillness, bliss and an absence of negative energy must be generated internally. No one else can do this for us. Powerful forces such as these can only be accessed after an unburdening of karmic debt along with a mental, spiritual and emotional house cleaning have been conducted.

Darkness must be emptied out to permit the entry of Light. Outdated personality traits, negative idiosyncrasies, unhealthy characteristics and outmoded patterns of behavior serve us no justice. Whenever you feel blockages and tensions arise from weaknesses such as these, immediately recognize, confront and release them. These hindrances are obstacles to consciousness expansion. Spiritual evolution requires letting the psychological demons go. Defeating our inner darkness is the truest form of exorcism. Successful personal transformation and spiritual evolution vitally require this asset in advance. We all have to let go of what doesn't work in order to create room for what does. Our negative traits are not born of Spirit. They are inherited symptoms of material existence. Defects are not a natural part of our divine quintessential construct. You will never know who you truly are if you spend a lifetime being someone that you are not. As infants, we are born innocent, not flawed. Thus it is imperative to reinforce the fact that negative conduct is learned behavior. It comes from the original programming that we receive as a child. It comes from environmental, social conditions, family atmosphere and many other causes outside of ourselves. It's not always our fault that they are there. But once we reach the age of maturity and reason, it is our fault if we choose to cling to these defects once they have been recognized and acknowledged. Personal self improvement is a human obligation. Individually generated positive vibrations contribute to the growth of the human race as a whole. Why? Because everything in creation is connected. That is fundamentally how Nature has been designed.

Despite the causes of our downfalls, the effects are always the same. Self defeat. We cannot transcend our awareness, connect to inner self or align with the collective consciousness of our Source with a heavy conscience. Our heart must never outweigh the sacred feather of Maat upon the scales. This creed ultimately applies to every day life in the land of the living, not just in the end in the realm in the Duat. When we cling to our defects all we are doing is lugging mental and emotional baggage around everywhere we go. All this serves to accomplish is to add even more baggage to the existing pile. It's an endless vicious cycle of insanity and a total gong show. So let it go. It is pointless to hang on to or cling to that which is self defeating. These types of choices will never result in mental, emotional or spiritual freedom from the frailties and vices of our biological ego self.

When using any modality of therapeutic energy healing system to detox unwanted energies, we must be willing to let go of whatever negativity arises. Releasing and letting go of the painful, unhealthy energy that we carry internally is the only way to make room for the good stuff. This is a critically imperative aspect of our personal growth on all levels. Healthier states of mind are the natural driving force behind making better choices. Healthy decisions promote proper human conduct that automatically aligns to Universal Law. States of being such as this contribute to a sense of inner harmony, feelings of connected oneness, unity, compassion and unconditional love. These forces align to the essence of the divine collective consciousness of our Creator. Connecting to this sacred omnipresent force through the gateway of the Inner Self grants us access to an information highway. The collective mind of the Vector Field is comprised of endless infinite knowledge. Within the collective consciousness resides the memories of every life experience generated by every soul who ever incarnated on the physical plane. In addition, the collective mind takes up residence in the plane of the Absolute, the vector field, wherein all other information ever perceived of by the mind of God exists.

In my own scrying practices the information that I tend to receive from the quantum field, the plane of the Absolute appears to operate by way of sympathetic resonance. The information feedback I receive from ascended spirits matches my own bio-field energy, conscious programming and harmonic frequency. In space and on earth, the Doppler effect measures both sound and light in the form of waves. Both of these energetic forces contain cosmic information. The higher consciousness that the mystic more or less naturally tunes into seems to correlate to and build upon whatever academic knowledge the practitioner has already attained. Even the most meager bits of good information that I have in regard to any subject of interest will be expanded on. As long as I am aware of something on some level, and understand the thing somewhat, my consciousness will attract further knowledge of *like nature.* All wisdom pertaining to all subjects known to humanity already exists in the collective mind in some form or another. Spirit doesn't send us or show us information which we cannot comprehend on any level. This waste of energy would be defined as a useless transmission. If information received from spirit is not understood, it cannot be properly recorded for use and further study in the material plane. This cause would only result in the loss of priceless, vital intelligence.

Communication of this sort is pointless, futile and irresponsible. Connecting to the collective consciousness will reveal to you, much more about what you already have learned or studied and properly comprehend even slightly. That appears to be what happens in my personal metaphysical practices. Consciousness is governed by the Laws of the Universe. The Law of Attraction is strongly at play here. As we know *Like attracts Like* so Richard Feynman strikes again. As a result of this interplay of sympathetic resonance we have the inherant ability to lock into information banks, the consciousness of Angelic beings, Saints and Ascended Masters who disperse spiritual knowledge. It appears that the wisdom they impart will align to our brainwave bio-field signature. As such, in regard to conscious contact with the collective mind the more you know, the more that spirit can tell you. In this way they can expand on any given subject of interest to you that resides within your awareness. That is how it seems to work. In a world where there is an unlimited amount of information surrounding this subject matter more people are beginning to experience spiritual awakenings similar to mine. Bearing this in mind, it becomes vitally obvious that acquiring as much good knowledge as possible is a spiritual obligation of every member of the human race.

Study of and exposure to a vast array of academic subjects is highly beneficial for anyone seeking spiritual evolution, alchemical transformation, conscious transcendence and a healthy, joyful existence in a higher vibrational state of being. With knowledge, comes freedom from mental vices. Humans are spiritual biological lifeforms equipped with cognitive rational thought capabilities and brains that enable us to absorb and store vast amounts information. So it is imperative to comprehend that the more knowledge we gain during our lifetime, the more information we can impart from spiritual contact with the Cosmic Mind. This guideline also applies to the astral world of prophetic dreaming. Wisdom is of vital significant relevance to human existence and evolution. The knowledge we embody individually and collectively as a race can be sensed by the spirit forces of Nature. It is known at the highest levels of the great collective mind of our Source reaching into the saintly realm of infinite universal truths. In this way, ascended masters, angelic beings or Saints such as St.Bernard de Clairveaux, Compte St.Germaine and so on, can potentially tap into your mental bio-frequency. When the human mind taps into the collective mind, volumes of information pertaining to subject matter with which you are already familiar, gets magnified, expanded upon, explained further and increased. As it enters your consciousness the knowledge flows through your internal stream of thoughts. Mediums use this channel to tap into various forms of disembodied human soul spirits vibrating at lower astral levels closer to the physical plane. The process of information retrieval can be very rapid. Thoughts must be recorded on paper through handwriting, imagery, artwork or symbolism in order to capture all information that is received. The flow of information presents as one continuous thought that can persist for quite some time. It doesn't stop, until it's done. In this way a authors who are capable of energetically connecting to the vector field can potentially write the words of Spirit through the ascended masters. There are several interesting books on the market that feature channeled information. The immaculate, divine energy of spirit is pure light, pure truth and pure wisdom.

It has been known to be the case that many great well known philosophical doctrines, religious manifestos, spiritual, biblical and esoteric literary works have already been produced for humanity in exactly this way. Clearly, knowledge is our greatest asset. Wisdom is applied knowledge. This asset is of even greater value than all of our material energy healing tools combined. Without wisdom, we wouldn't even know what tools to choose let alone how to design or utilize them. In the eyes of our Source, light and wisdom unite as one divine force. These two forces in combination are necessary requirements for achieving transcended awareness, expansion of consciousness, personal growth and self mastery. Any mystical experience we can ever hope to achieve by way of controlled, directed, conscious intent is rooted in a clear, sharp, mental state, keen perception and the ability to focus. The achievement of a higher vibratory state of bio-field resonance automatically elevates our state of consciousness. Generating such states of existence are key to expanding our energy beyond the confines of our biological earth vessels and objective mind. There are many forms of harmonic resonance healing tools available on the market that promote mental expansion and transcendence. Referenced in the final two chapters of this book is valuable information regarding bio-field technology energy healing and brainwave entrainment achieved through the power of the mind at will.

In 2021 I discovered and began utilizing the most highly advanced bio-field technology program available on today's global market. This system promises to induce a profound spiritual awakening. That is exactly what it has done in my case. When properly utilized and applied as directed, transcendent states of being can absolutely be achieved using this mind blowing bio-tech full spectrum meditation series. This outstanding, highly advanced bio-field technology has helped me deconstruct everything within my body, mind and spirit that I didn't need, wasn't healthy and was not working. It further gave me the ability to reconstruct an entirely new, much better version of myself. Energetically, a newer more healthier attuned model, Marika Sophis 3.0. Life becomes much less painful when we choose to live in a higher state of loving, positive vibration. In this way, we attune our energy to the *Oneness* and healing energy of the Spirit Forces of Nature existing all around us. We align our energy to our planet, to our Source and to all other members of humanity as well. What I am sharing with you in this book, is a very powerful, uniquely designed energy healing system that is in a class of its own. In the two final chapters of this book, I offer you my personal testimonial and the effects produced through this healing system after 1 full year of use. I tell you what it is, where to find it, how much it costs and why it's worth every penny, even if you have to budget to buy this program. I am pleased to share with you, what it did for me and the spiritual levels of development I have attained so far.

Utilizing this advanced binaural meditation program actually produced visible, observable, evident positive effects and changes in regard to my personality and sense of self. Working with this full spectrum bio-field technology encouraged and generated release of all energetic obstacles and blockages. This system facilitated tremendous growth and improvement of my overall sense of being. Bio-field technology of this intense nature has endless, immeasurable physical and spiritual benefits. It offers numerous healing rewards and produces tangible results that you can actually see, sense and feel. I can most certainly tell you that this system has helped me change my entire state of mind, emotion and sense of being all for the better. These healthy improvements benefit myself and also extend to the greater good of all those around me. It has helped me create a mindset that automatically aligns to the codes and standards of Universal Law. This is exactly what the human spirit, physical body and consciousness are supposed to do quite naturally. This is how humanity has been designed. If we are all to become united nations of joyous citizens, co-operatively, gladly and willingly abiding by Natural Law, certain spiritual requirements must be met. Many of us do wish to achieve ultimate goals such as saving our planet of our own free will. Furthermore we all desire to live in healthy stable communities. The world is a mess and we all know it. Humanity knows that our race and planet are currently in decline. What will we do about it? How do we create *Order out of Chaos*?

Harmony and order require balance. All aspects of Nature, including faithful spiritual people devoted to various paths of light and worship are necessary to the growth of our race as whole. Nothing in creation is an accident and all things are set in place for a reason. Faith and Hope are significantly imperative human energies to embody, grow and foster. These forces are essentially good. All loving, ethical religions, paths of light, spiritual traditions, esoteric faiths and so on, are roads that eventually lead to the same Source. We needn't agree with all of them. Our spiritual obligation is to accept people for who they are regardless of what they believe in. The Absolute gave each of us free will to choose our own spiritual paths for ourselves. Not the right to judge others for the path they choose.

 In regard to paths to light and faith, humanity simply has many options. None of them are 100% correct, nor are they 100% incorrect. God is the same God no matter name we attach or do not attach to our Creator based on cultural and ethnic differences. All religions and faiths centered upon a just, loving God have one common thread. A powerful creed relating to proper ethical human behavior and conduct as spiritual, biological aspects of Nature. This, is the really good thing that all positive, loving, heart centered faiths and belief systems all have in common. Without this, there would be no spiritual order within our race and no ethical method by which to establish it. Regardless of all of this, Jesus is good. Mohammed is good. Buddha is good. You see? All that is of a loving God, is from God. Faith is good. Devotion to Universal spirit in some form suitable to you, is always good. Oneness created by true unblemished Christian unity is also very good energy to send outwards into the world of living things. Christ consciousness relates to our internal connection to our Source which is the ultimate light. Regardless of our religious preferences, this divine, internal light can be discovered and experienced by anyone. Religious and spiritual paths are simply a tool that serve to guide faithful, light seeking pilgrims along on their way. This is so, for the simple fact that the illuminated light of Christ consciousness refers to an enlightened state of being that Christ himself attained while he was here with us on earth. Indeed he absolutely did come here. Yes. He was a highly ascended soul who loved humanity and God enough to incarnate as biological flesh, a member of the human race. His life mission really wasn't about all the creeds and laws of the church that many of us may agree or disagree with. His main purpose was to show humanity the type of conduct, ethics, choices and lifestyle that would draw each of us ever closer to our Inner Selves and our Divine Source. He came to show us how to act proper in the eyes of our Creator. He loved all equally. He gave to the poor. He healed the sick. He did this, like a Super Star.

If Jesus were here today perhaps the current state of the church would concern him. From another perspective, in his perfect glory he may see even farther into the issue and think it needs a complete overall and reformation. Christianity teaches us that he died for our sins. Universal Law teaches us that we ourselves, not Divinity are responsible for our own shortcomings. I'm not going to attempt to resolve this confusing contradiction. He was actually here to help. Look what was done to him. The work of dying for the sins of humanity has been successfully conducted. So be it. The deed is done. Its time to let all people everywhere bless him in return for his teachings, sacrifices, love and affection for us. Let us rejuvenate his spirit within our hearts and offer him roses. It may be benefit our race to allow his angels to cleanse his ancient wounds. It will also benefit our race to remember his greatest sacrifice. Which is that he descended from a great, lofty angelic height to come here as a man. Thus, he was not perfect. But he was as close to perfect as it gets. Can we not thank him, relieve him of his gory vocation upon he cross respect his true teachings and move forward as responsible aspects of Nature? Is this too much to ask? Most likely.

Humanity is *One* living, breathing, interconnected biological lifeform. This divine unity is the glue that binds us all together. In order to achieve global unity, we must eventually unite the spiritual cultures of the East and West. There should be no issue with this in regard to spirituality but there always has been. This is very ironic when one considers that in reality, we are One. In the first book of this series I explained the Seed of Life as it correlates to the work of creation scientifically conducted during the 7 days of Genesis. The pine cone statue in the Vatican courtyard is a symbol of eastern philosophy. Are Christians capable of seeking answers to unanswered spiritual questions elsewhere? When the bible does not provide certain necessary explanations can science fill the gap? Yes, it can. In addition, taking God out of the equation is not a pre-requisite to embracing the sacred nature of science.

What does the bible teach us? That our Source created the Universe in 7 days. It does not offer a full detailed, description of the scientific method. What does the Seed of Life describe? The scientific process explaining how our Creator manifested all things in existence in those 7 days. Combined, these two forces shed light upon one another in the most brilliant way. Together, they offer a detailed account of how the scientific processes of Nature contribute to the intelligently designed manifestations of the Absolute. Science is the work of God. The Seed of Life is at the heart of the creative process. Without it nothing would be here. There would be no way to explain fractals, sacred geometric archetypes, Fibonacci, the spectrum of light or the fabric of the vector field. Metatron's Cube crystallizes energy into the matter required to form our bodies. Consciousness comes from our Source. Who created Metatron's cube? The mind of God. It is the work of our Creator. These are inseparable forces. Ancient philosophers of the East have always known that the Seed explains the Genesis. To become a global community of *oneness* bound by unconditional love, the division between the eastern and western schools of thought must also be *one*. This is logical to assume. Deep devotional, intellectual faith has been known to help cure many spiritual issues that plague the human race.

When I see the Seed of Life I know that I am be-holding the Scientific work of God-Mind in all its glory. Could this knowledge aid Christians in developing a better understanding of the Biblical texts? What about the meaning behind the two breath taking Bramante Spiral staircases in the Vatican museum? I think it would answer many questions. The problem has nothing to do with our ability to develop faith. The catalyst to blame in part for the obvious global decline in church attendance and loss of interest in religion may boil down to logic. Blind faith does very little to satisfy todays inquiring minds. Many religions are rooted in the vaguely presented, incomplete information which leads to unanswered questions and states of confusion. This fact is commonly known. How do we remedy this? With proper knowledge and intelligent, rational information. Why is any of this even important? Because when members of the human race express faith through prayer, devotion, love for our Creator and towards one another, this cause raises the harmonic frequency and elevates the sympathetic vibration of humanity as a whole. Why? This effect occurs as a result of the omnipresent connectedness of all things. In metaphysics this truth is firmly established as one of the most fundamental aspects of Nature. All is *one*.

The energy being raised in New Age spiritual practices and traditions works in the same way. Faith in our Creator and in the spirit forces of Nature is an essential healing energy. Without it, humanity would always be dwelling in negative mindsets with no path of escape. If you have faith in positive, loving aspects of Nature and exhibit selfless compassion towards others it benefits all of us. From this perspective, is it not profoundly obvious that no member of humanity has a right to judge another for their choice of faith? Yes it is. Furthermore, do we have any right to judge their mindful forms of devotion or the name they assign to the mind of God? No. In fact, our Creator does indeed answer to many names. This is so. At the end of the day, is the Catholic church uneducated in the ways of the East? No. But these two schools of thought will always clash. They are opposing platforms offering two separate systems of belief. Currently, one is rising in popularity while the other is waning. One is in need of re-structure. These spiritual forces in proper combination, could potentially change the face of Christianity for the better. In the Aquarian age of reason, it's a lovely concept to consider. Sadly, it's also an outrageous thought and a highly unlikely probability. Regardless, in my open minded research, Eastern faiths and science have both shed light on unanswered questions in the Bible. These paths fundamentally support each other in theory, philosophy, historical accounts and various metaphysical truths. Do the research, you may reach the same conclusion. Combined, these pools of knowledge produce a force more powerful than the impact of each sole individual component on its own. Ultimately, faithful parishioners are looking for the same fundamental answers that new age or eastern philosophical souls are. The only difference is that they seek their Light elsewhere. The answers to many spiritual questions can indeed be answered by science. The philosophical questions that the Bible raises which remain unanswered by the Church, can be answered by science and in many cases, scientifically proven. The Truth has always been the *Solution* and the *Way* to the Light of our Creator.

Isotropy, consistency and uniformity of biological lifeforms in Nature is accepted proven science. The existence of fractals, Fibonacci, the Golden Ratio, platonic solids and sacred geometric archetypes are known accepted aspects of metaphysical creation. We know that the Universe and our bodies are electric by natural construct. This science has been and is still being proven. The evidence abounds. New technology is providing further evidence by the day to help us examine this concept. The James Webb Space Telescope is in the ultimate position to show humanity galaxy-quasar interactions up close. With this evidence, modern science can either prove or disprove the potential true nature of redshift, look back time and quasar origins. Vital knowledge such as this is of unlimited value in regard to identifying quintessential scientific correlations to Natural Law.

It is human nature to dream of a Utopia where the world is one, unified illuminated force with no crime and no poverty. How do we generate a result such as this? How can we make this happen? Do we have what it takes to create a better earth and future for the next human generation? What will we leave behind for the youth of the nations that we live in? Causes such as these, generate significant reasons to care about our planetary home and the condition of it. Positive effects result from positive action in regard to tackling obvious environmental issues. The problems that are causing the rapid destruction of earth and the overwhelming amount of damage and pollution, stem from the living habits of the human race. Its time to save the planet for the greater good of all. Not because humanity supports or does not support climate change. Not in fear of penalties or fines, but in willing co-operation of our own desire because we actually care. In reality, we do not even own this planet. Everything we see all around us, everywhere in Nature, is on loan to us. Just as it has been for every generation of humanity who has ever lived before us and those who shall ever succeed us. Thus, no member of humanity has a right to destroy it. Yet, look at the state of our earth and the damage we have done.

As intellectual, biological aspects of Nature we are all responsible for proper stewardship of this planet while we are here. If Mother Earth is in terrible condition, low vibration and poor health all biological life on the surface feels this energy too. It is our responsibility as a whole to fix and remedy this disempowering depleted energetic state. Unhealthy energy creates obstacles to proper biological and spiritual cosmic attunement. From holes in the ozone layer, to the trash heap in the sea the damage to earth has been assessed by spirit. We have an alarming critical dilemma to attend to as a race. Humanity has done an incredible amount of careless damage to something borrowed. Should we focus our energy on saving the planet? Yes indeed. It is our biological and spiritual obligation. Each and every one of us must make a personal effort to do this, interdependently, together as *One.* Our home is a mystical living being, as are we. We are from earth and our biological composition is of earth and returns to earth. It is now the dawn of the new age of reason. Perhaps the time has finally come to vindicate the teachings of great philosophers such as Solon, Pythagoras, Aristotle, Harvey Spencer Lewis and Rudolph Steiner along with other wise scientific masters of this current age of humanity. For those who seek to truly be of assistance in healing the planet, I recommend a study the powerful science of Bio-Geometry patented by Ibrahim Karim of Cairo Egypt. That is to be my next magnificently empowering, greatly illuminating academic endeavor.

JWST, equipped with its 48.25 grams of 24 karat gold, hexagonal imaging chips is out there viewing the Cosmos in infrared. The mysteries of the Universe can only stay hidden for so long. Correlations between Universal Law, the construct of the Cosmic Egg, the Seed of Life origin theory and the electric universe model are endless. Evidence of this Nature abounds all throughout creation. On the other hand, the Big Bang explosion was a one time random event. It cannot be witnessed repeating itself in biological Nature in any observable way in our biological, material world. Many professional scientists and common people alike are fully aware of this scientific discrepancy concerning this standard model teaching. More evidence surfaces daily in support of new theoretical creation models and the abundance of electricity in our cosmos. Upgrading current scientific modes of thinking along with an evolution of certain standards and their limiting mental confines is inevitable. This change is soon to come. The scientific revolution and the spiritual revolution are one thing. One event. Two forces combined into one force united.

The youth of our Nations, the next generation are sharp, bright, very inquisitive and incredibly smart. Put to the test in a world where brilliant new forms of technology are on the rise, their minds are likely to excel quite impressively. Outdated scientific standard models that no longer work in light of new emerging evidence, are not likely to satisfy or appeal to their clever curiosities. Scientific evidence mounts in support of alternative origin theories, alternative redshift and gravity theories. On the same token, this evidence is working against the concept an expanding universe. It is becoming more scientifically obvious that the universe we live in is of electrical construct. This new insight may cause the need for a re-evaluation of what we think we know about Science. Standard models of science that no longer work may find it difficult to stand up to todays technology and what it can reveal to humanity. The current standard models of science are also not necessarily in sync with the growing pool of knowledge being generated by the modern New Age spiritual community. Many members of the human race can sense an academic misalignment associated with contradicting scientific views.

Creation theories and scientific processes of Nature that align to Natural Law show the quintessential connections between science and spirituality. Endlessly recurring patterns of creation and manifestation that align to universal law can be witnessed repeating themselves all throughout Nature. Many intelligent people have no interest in studying science due to contradictions between questionable scientific models and revolutionary new, ground breaking discoveries. Another catalyst that generates disinterest is the fact that physics has been overcomplicated with endless, unnecessary mathematical equations that can fill 3 classroom chalk boards. My goal is to simplify the science as much as possible to generate an interest in learning valuable information that is relevant to our existence and our purpose as a species. In this book, we will explore an alternative model of galaxy formation generated by the Galileo of Palomar, Astrologer Halton Arp. Will we do this with an open, objective mind while also considering the basic fundamental aspects of known, standard model paradigms. Looking at both sides of the story and examining both schools of thought promotes rational, intelligent mental expansion and a good sense of reasoning. The Scientific community is currently divided between standard models of science and mounting evidence of electricity found everywhere in the Universe. Intelligent, alternative theories regarding redshift, galaxy formation, quasar origins, functions of gravity, the expanding universe and so on, do exist. These rational logical, heavily researched and properly evidenced concepts are all piling up on the scales of reason. It is clearly evident that we are on the brink of a massive change. Meanwhile, the New Age spiritual society is exploding with teachings of metaphysics, sacred harmonics, sacred geometry, energy healing modalities and sympathetic frequencies. I cannot stress enough that the Scientific revolution and the Spiritual revolution are *One* Solution. They are the same event. The new spiritual way of the Aquarian age of light and reason is the emergence of *One* global tradition comprised of these two powerful forces in sympathetic combination. A compassionate, enlightened society bound by unity, unconditional love, obediently operating via the codes of conduct set forth by Universal Law, is the optimum. An excellent goal to strive for. This is a superior mode of operation in comparison to a anxious society bound by forces of stress, fear, lack, punishment and limitation.

In times of great change, willing co-operation is ideal. It renders no consequences and does not limit free will. Holding the truth hostage serves no justice to anyone. To develop a better understanding of our existence, humanity would require only a crumb of validation that proves the electric universe theories which are already so heavily evidenced. This acceptance and acknowledgement must come from the standard model Scientific community. There is also plenty of proof available that provides a solid foundation in support of an origin theory that resonates with and corresponds to Natural Law. In combination, the metaphysical theory of the Cosmic Egg and the Holographic Universe do align to observable, orderly scientific processes of nature. This can be witnessed within the great macrocosmic to the most minuscule aspects of creation. Through developing a foundational scientific knowledge of Natural Law, humanity will inevitably reach the conclusion that we are not alone in the Universe. In fact the law of correspondence teaches us that it is scientifically impossible for our human race to be anything more than a model in Nature. An educated society with a knowledge of fractals, sacred geometry, quantum physics, numerology, Fibonacci, the golden ratio would know that we are not alone. With this understanding they would see and comprehend the great patterned structure of Nature. Demystifying spiritual aspects of science gives all of humanity a reason to look up to the stars in awe and wonder, together as *One.* Establishing unity as a race is a pre-requisite to achieving spiritual evolution. Are we awakened enough to meet the standards of an age of reason and light? In this new age, illumination will come more easily to those capable of receiving the light, so we shall see. We are all witnessing the great cyclical work of our Source in action. It is, always has been and always will be, immaculate to behold.

To become a part of this new age of light, one requirement is a willing desire to become the very best version of ourselves that we possibly can be. We must strive to achieve this elevated state of existence for the unconditional love and greater good of all. Energy healing tools such as the bio-field technology referenced in this book are excellent aids that promote spiritual development. Harmonic frequencies and binaural track sessions recommended in my 1st two books can certainly help you along on your way to attaining higher states of conciousness. In the final chapter of this book I share a method of how to generate instant brainwave entrainment and slow heart rate at will. This priceless, simple technique can be performed by anyone, anytime, anywhere. The most fundamental aspect of this process is an excellent sense of memory recall along with the ability to concentrate and focus. Through mastering certain mental faculties we are able to achieve transcendent states, calm our emotions, decrease physical pain or ground and center our energy whenever necessary. Utilizing this method is easy and requires no props. It can be used at any time of day, whenever the need arises. In the final chapter of this book we examine this method generously shared with me by Energy Healer, Music Producer, Published Author, Dameon Keller. Let's now move on and take a reflective look into Universal Law as a creative force and binding agent for all things in existence.

CHAPTER 1

The Oneness Of Universal Law. The Glue That Binds All Life Together

There are 7 principles of Natural Law that govern and regulate all aspects of creation. In the 1st book of this series *Macrocosmic Reflections in a Microcosmic Mirror* we looked at a few of these laws and how they correlate to our daily lives. We explored the Law of Attraction, the Law of Correspondence, the Law of Duality, the Law of Cause and Effect and the Force of Karma. Ultimately researching the Universal Laws of creation endlessly shows us how everything connects. All is One. Natural Law is the omnipresent glue that will always bind all things in the physical plane together. Everything that exists in our biological material world is a reflection of greater creative processes occurring elsewhere in the Cosmos. All things in creation sprang to life from the mind of our Source. Metaphysically speaking, this truth is infinitely encoded within the principle or Law of Mentalism.

The Law of Mentalism teaches us that everything in existence within the great manifestation, our holographic universe that we live in has emerged from the consciousness of the Absolute. In Eastern philosophy our Creator utters the word AUM and the seed of life forms. In the Bible God says *Let there be Light* and the eye of the Vesica Pisces opens wide and blasts a jet stream of ultraviolet electromagnetic plasma light outwards. This light eventually fills the physical plane as it moves through the womb of creation. In metaphysics and Alchemy this is also what is known to be occurring. A word or logos is uttered, a Vibration is sent forth. It travels through the vector field like ripples in a pond through the vast electromagnetic plasma sea. All of this begins with the mind of our Source enclosing part of its consciousness within a spherical energetic barrier. Herein the Divine Consciousness takes up residence in the perfect position within the sphere as the Immaculate center. God as such, becomes universally symbolized as the well known geometric Archetype, a perfect circle with a dot in the middle. Essentially, this is the most magnificent quintessential symbol known to humanity. The Law of Mentalism is firmly structured upon this symbolism. Thus, in regard to understanding Universal Law this knowledge is of tremendous value. The Circle with the dot in the center has been the most commonly known symbol representing our Source since the dawn of civilization. Humanity also relates this symbol to another divine, creative, life giving force in Nature, our Sun. Without it nothing would be born on earth. Biological life could not survive or exist.

Consequently from this primordial archetype, the Divine immaculate Sphere, all things in existence will begin to take shape. The circle with the dot in the middle is the foundation stone of the Seed of Life. From the Seed comes the Fruit of Life. From this sacred geometric architecture, Metatrons Cube emerges. Finally, through the guided consciousness of Divinity the creative force evolves into the Isotropic Vector Matrix with the tesseract and vector equilibrium at its heart. From the sacred cosmic construction of Metatrons Cube, all biological life begins to manifest through the process of crystallization of Light energy into matter. The driving forces behind this great work are fractals, the golden ratio, Fibonacci sequences and spiraling, helical, electromagnetic Birkland currents. This is the glorious creative work that was conducted. It came forth from the radiant mind and consciousness of our omnipresent Source during the 7 creative days of Genesis. This knowledge has been pre-programmed and etched into the memories, hearts and DNA of humanity. This has been the case throughout every generation and civilization of our species. As a race, we already know these things. All humanity truly needs is a reminder and a wake up call. This is the way. It has always been the way and it will always be the way to the light. Amen.

Obviously the consciousness of the Absolute is omnipresent. This means that it can be found everywhere all throughout creation. From the galaxies to an insect standing on earth, everything is connected through the same primordial patterns of Creation. Manifestation began at the dawn of Genesis with a circle and a dot. This information has belonged to humanity since long before Constantine ruled Rome. Before Summerian cites were raised to mighty civilizations. Before the Maya and the Aztecs walked North American plains. Long before the pyramids were ever built. Even before the Sphinx was built. This is a primordial construct of Nature, a model of creation that pre-dates all others and this information is encoded in our DNA. This wisdom is still alive today. It is woven into the fabric of all metaphysical esoteric faiths and Eastern philosophies. It has indeed been known for ages and handed down through many generations of ascended masters, brilliant philosophers, ancient tribes, ancient astronomers, sages, saints and saviors. This is so. The Law of Mentalism demonstrates the glorious, immaculate power of the mind of God. The Flower of Life, further explains the Genesis of creation. Christianity deserves to benefit from this knowledge. In this way they can study and learn the Bible with fresh new eyes of Light. In a perfect world, this would come to pass.

Everything we see in Nature, in the cosmos and in our Holographic Universe all around us exists by way of the divinely directed orderly, controlled thought of our Source. Grasping this key factor and understanding that humans have been made in the image of our Source helps us understand how a creative power similar to this also exists within the consciousness of humans. We too, have the ability to mentally manifest our reality. Book one of this series *Macrocosmic Reflections in a Microcosmic Mirror* delved deeply into this reflective metaphysical paradigm. Our Source, is ultimately the profound, powerful, omnipresent, formless energy of divine consciousness. God is essentially an immaculate mind. A colossal invisible mind. Metaphorically speaking, a big giant head, that is slightly similar to the concept portrayed in the popular sitcom 3rd Rock from the Sun or the wizard of Oz. From a metaphysical perspective, our Source, the Absolute, is ultimately an intelligent energy, with infinite creative potential found everywhere, dispersed evenly all throughout the Vector Field. In addition, this divine consciousness is composed of blinding radiant golden light and endless, infinite unconditional love. Everything that our Source manifests in the electromagnetic plasma sea of the vector field, it creates through love and curiosity. Our Creator has metaphysically constructed and devised several harmonic, vibratory levels of geometric existence of different densities to promote human and angelic soul development. God is a scientist, an expert quantum physicist. Our souls stem from the same source as those of the spirit beings and principle energies that reside within the cosmic and physical planes. These are the entities or beings that are participating in the great repetitive system of reincarnation. As humans, our souls are a part of that process. A study of Eastern philosophy and mystical science can facilitate a deeper understanding of this profound, sacred scientific truth. When our souls are fully developed and purified through multiple incarnations, we return home to our Source and the information, our programming, all that we have learned in every lifetime, gets added to the vast knowledge database of the collective consciousness of God-Mind in the plane of the Absolute. This vast pool of endless, infinite knowledge along with the consciousness of those existing in both the physical and cosmic planes, appears to be exactly what psychics, empaths, channelers and mediums have the ability to tap into.

Thought generates manifestation in the creative work of our Source. In comparison, our own thoughts shape our choices, actions, deeds, ethics, conduct and so on. Through our own consciousness we create the experiences that we draw forth into our lives. These experiences manifest into our own personal version of biological existence and reality. Why am I writing these 4 books back to back during the Chinese Zodiac, 2023 the year of the rabbit? The rabbit is here to heal us. Last year, the year of the Tiger, ushered in plenty of angry, chaotic, negative energy for many. The tiger did indeed attempt to tear some peoples livelihoods to shreds. The healing energy of the rabbit was an excellent successor. It's here to rejuvenate us. Not hurl us into another rabbit hole. I'm writing to humanity because I care about the state of our world. I feel unconditional love toward the entirety of our human race and also compassion for those less fortunate than others. I feel this way despite our flawed nature, that is in many ways in a current state of rapid spiritual decline. If any of this information leads anyone to their own light of truth, my efforts will not be in vain. That being said, this message may not appeal to every member of humanity. If so, so be it. It will appeal to those seeking *Light and Truth* that wish to contribute to the Spiritual evolution of the human race. Despite that stipulation, this message is and always will be, for everyone on earth. After all the losses so many all around the world have suffered in the last two years, I see this depository of scientific knowledge as my selfless contribution to help heal the many lost or somewhat broken, exhausted, disempowered members of our human race. Let us not lose hope in a joyous, harmonious future. There are far too many other options and absolutely no need of that.

Discovering how to create and maintain healthy energetic states of positively harmonized energy and a balanced body, mind and spirit is vital. One key to achieving states of existence such as this is to develop an understanding of the Law of Vibration and how it correlates to our overall sense of being. In my 1st book *Awaken and Ascend* we explored the advantages, benefits and disadvantages of being in higher or lower states of vibratory frequency. Energy healing tools and binaurals were suggested. Utilization of these frequencies facilitates alignment to Inner Self, to the Spirit forces of Nature and to higher energetic vibratory states. These changes take place on all levels emotionally, spiritually and physically. Understanding the Law of Vibration helps us comprehend the nature of our universe. The first thing that the Law of Vibration teaches us, is that nothing in our material world is actually really here. This is an immense, unfathomable concept for the average person who does not study quantum physics, to wrap their head around or embrace. To many people, this type of logic doesn't even make any rational sense. We all know that when we touch a table or a rock it feels hard and impenetrable. It appears to be solid. When we walk on the earth, we don't fall through it. When we get into a vehicle, we are surrounded by a few tons of steel and it appears to provide some form of protection. Furthermore most people do believe that their body is composed of solid matter. All of these observations do repeatedly present a good case that supports belief of a so-called solid world. Yet, in great contrast, nothing could be further from the truth. The hollow construct of our physical world and the cosmos all around us has already been proven by science. As always, science comes to the rescue.

In the world of science, observable evidence proving our hollow existence and our equally hollow world is beyond abundant. You don't even have to be a professional scientist or a Harvard graduate to see it. Any intelligent person with an interest in biology or chemistry and access to a microscope can see it for themselves. Everything in our universe is composed of atoms, molecules, elements and minerals. Neutrons, electrons and protons metaphysically latticed together comprise the atoms and molecules that construct everything we see in existence, including us. It is a well known scientific fact that there is nothing more hollow than an atom or a molecule. This is exactly what we see when we look at biological cells through the lens of a microscope. It's that simple. There are a whole host of reasons why everything appears and feels solid to us. Once again we can look to science to find these answers. Understanding the scientific processes of Nature helps us embrace the expansive wisdom that explains our hollow world. There is a vast variety and great diverse volume of different types of manifested lifeforms in existence. The main, vital difference between all these distinct life forms is the varying rates speed which the light they are composed of vibrates.

Each perfectly manifested thing in creation has its own sympathetic frequency that acts as an energy stamp or bio-field signature. Each particular thing in Nature is encoded in this way all throughout creation.This is where our false sense of individuality, the illusion of division stems from. The structure of an atom reflects the structure of a solar system which in turn reflects the construct of great galaxies. At the center of all of these powerful forces, we find an electric nucleus. A pulsating plasmoid. In the electric universe theory, this electrical phenomenon takes super massive black holes right out of the picture. In the microscopic aspects of biological life, the grand structural pattern of primordial creation repeats itself in the model of a cell with its powerful nucleus. The same holds true for the proposed architecture of the cosmic egg. Theoretically, in certain scientific and metaphysical schools of thought, it's structure is thought to resemble a giant cell like organism with a powerful nucleus that generates our dual holographic universe. From this, came all other manifestations within the matrix of life. These various forms of biological life emerge as an orderly reflective process of this original, initial macrocosmic manifestation. This is how Nature works. This wisdom provides an understanding of the work of our Creator. Each and every material manifestation is comprised of Light energy. Essentially, electromagnetism and plasma in a constant state of vibration.

The human body, our biological earth vessel, maintains its constant state of vibratory resonance, until our soul leaves our body to go to the Cosmic Plane. This is where we rest and contemplate the lessons learned in the life we've lived and what is needed for soul development in the lifetimes still to come. When our life force leaves our body, the biological organism responds by going into a state of decay. The vibrational state of our biological earth vessel now changes to a very low state. Once inanimate, the body is without a soul. In Nature and metaphysics, the body is not considered to be dead. This is simply a stage of biological de-evolution. Inside the body, the physical cells continue to break down into minerals and elements that are recycled back into Nature and the earth from whence it came. The energy that once comprised a human body, now becomes another part of Nature. The cells, while in this state are not technically dead in the way that we perceive death. In fact there is still a tremendous amount of active vibrational processes operating during this physical cellular and molecular decomposition. Dust to dust as it says in the Bible. This statement tells us that our bodies are of Nature and that they return to Nature. Our soul and consciousness are designed to return home to the realm of God. This only occurs after many necessary incarnations and pit stops in the Cosmic Plane have been completed. That's what we know to be true in metaphysics and how it is viewed from an alchemical perspective.

Consequently, when our Souls return home to God, what is really happening metaphysically is that our spirit force and consciousness are returning to the plane of the Absolute. The immortal vector field. The great vast sea of golden waves of electromagnetism and infinite potential. Herein we find oceans of electromagnetic plasma currents of highly charged, undulating transverse waves taking on helical form and great spin following universal laws, fractal equations and golden spirals. From this Fibonacci and the Golden Ratio, sacred archetypes and platonic solids are born. Emergence of more complex energy forms such as Metatrons Cube can now occur. This is what is going on inside the Mind of God. The vector field is filled with intelligent, orderly consciousness of the immaculate mind of our Source. In this realm, the Absolute has a thought and before you know it, a universe comes into existence that is teaming with life. Energy within it is dispersed everywhere in various states of vibration and harmonic frequency. These energy forms sympathetically resonate as one divine, immaculately conceived manifestation. All biological things within this world of life, dance together to the song of C scientific pitch and A432Hz while radiating the unconditional love of the Absolute.

Light comes to life through the consciousness of our Creator. We as humans are a part of this glorious symphonic orchestra of light and life. If you do not understand this correlation, reading my 1st two books will facilitate your comprehension it. This, is very cool science. So, as all things are comprised of light in a state of vibration, everything is here, but also it is not here. The profound mind of God is quintessentially immaculate beyond the scope of our limited human understanding. We are to revere this force of consciousness from whence we have emerged. We are to display tremendous gratitude towards this divine alchemical force for our existence. For the air we breathe, for this gift of cognitive consciousness, rationality, logic the power of free will and our internal moral compass. The spirit forces of Nature and our earth have shaped our human heart. The unconditional love of our Source flows through it in the form of light. God, our Creator, by any name you choose to call it, is good. All the time. That's a fact. Christians are not wrong about that. Nor are the Muslims or the Alchemists, the Mystics, or the New Age Spiritual Traditions and all other humans who also respect and honor this creed.

As stated multiple times but never enough, we are *One*. Natural Law shows us this over and over again. Once you come to understand the scientific processes of Nature you reach further back in time into ancient knowledge.What we tend to discover is that corresponding reflective elements between biological manifestations exist everywhere. Like atoms and molecules that bind all physical matter together, everything around us is hollow energy in a state of vibration. This energy in its most primitive state, is Light itself. From this perspective, everything we see that we think is solid matter, is really light in numerous states of form and vibrational existence. All is composed of hollow atoms and molecules, in a constant state of motion that can be easily observed and proven using a microscope. Mentally embracing this profound concept facilitates comprehension and acceptance of the archetypal structure our holographic universe. The entire universe as well as the scientific processes and systems operating within it are an identical reflection of how our bodies are structured and how they function. We begin these comparisons at zero point creation with the fact that the Cosmic Egg itself is ultimately the original, fundamental model of the cell. The Cosmic Egg is floating in a highly charged electromagnetic plasma sea. Just as our biological cells are. How profound and extraordinary is that?

Our biological earth vessels, our human bodies, are comprised 70% water filled with electrically charged molecular atoms floating around in it. The water contains infinite numbers of cells always in a state of growth, development and regeneration. These cells, are of like nature to the cosmic egg which contains a very powerful nucleus. Our bodies also house the profound *mental* force of consciousness within us. This is the power of our Creator that we embody internally. It has an automatic connection to Nature and is capable of Inner Self alignment at all times. We absorb the *spirit* force of our Creator through breathing clean fresh air, absorbing ultraviolet and solar radiation, sunlight. If we are to look to the functions of electricity in our universe we come to comprehend how we absorb powerful electromagnetic currents from the earth. We further absorb the healing forces of Nature in the form of naturally moving water, lightning and rain storms, hot springs and so on. Every one of these aspects of creation is connected to the divine omnipresent Source energy at all times. To connect with Nature is to connect to Spirit, to our Creator, the Absolute, God. Everything in existence connects when we examine creation and the Law of Correspondence through the model of the cosmic egg and holographic universe theory. These origin theories also rationally correspond to the functions of our electric universe. In debatable contrast, the Big Bang theory quite simply does not. This poses a scientific paradox.

In regard to the Big Bang concept of creation, I have found no apparent parallels to this model in existence anywhere in Nature. It is simple to see that this raises alarming issues in relation to examining an origin theory that sympathetically resonates with Universal Law. Knowledge of the biological functions of our bodies and the things we see in Nature all around us give us insights into the greater macrocosmic aspects of creation that are occurring at cosmic levels. These greater processes include metaphysical functions such as the birth of stars, galaxy formation and galactic quasar ejections. All of the alchemical, quintessential correlations must be acknowledged, researched, proven, accepted, studied and comprehended if we ever wish to make any sense of our existence. Christianity is right. Our Source created all things. The pro-creation scientists are right, the universe is of very intelligent design. The Bible is right. It teaches us that all things were created in 7 days from the mind and harmonic utterances of God. Eastern Philosophical Traditions are right in explaining how this divine manifestation work was conducted through the construction of the Seed of Life. Each ethical religion and spiritual path on earth that is based in goodness and love contains some version of a set of rules or guidelines governing proper, moral human conduct. Therein lies the true common thread. This is one of the reasons that all ethical roads to God lead us all to the same ultimate Immaculate Source energy. So why are we so busy judging each other based on spiritual beliefs? Who even gave us a right to that? Human conduct such as this most certainly does not align to Natural Law. Be it the 10 Commandments, the basic ethics outlined in the Koran, Universal or Nat ural Law, the Golden Rule and so on, all of these set standard codes of moral conduct produce the same beneficial results. So the bottom line when it comes to the spiritual paths that others follow, judge no one for what they believe or have faith in. For the love of God, just have faith in something positive loving and good that appeals to you. Even if you find it in Egypt, Mecca or Persia or India and you happen to be an average Caucasian North American person. It's irrelevant where you find the Light of God. What is imperative is that you find it somewhere. Our human race has devised and produced many good systems of beliefs, creeds, religions, spiritual and new age traditions which essentially lead all pilgrims back to same Creator. The one and only Absolute omnipresent, divine mind and consciousness of our one true Source. In regard to heavenly abodes and returning to God when we die, Christianity is correct. This does indeed occur but only after the spiritual, evolutionary Soul work of the East has been conducted. For both teachings bear truth, we must also incarnate many times before we can go home. This is how our Souls become purified enough to be in the immaculate, radiant glory, essence and presence of the Absolute, our Source. This is how our consciousness evolves and acquires the essential wisdom and experience necessary to contribute to the vast amount of knowledge already existing within the great information memory bank known to mystics as the universal Collective Mind. It takes many lifetimes and incarnate experiences to annihilate all weaknesses from our Soul. Only then can we have established enough wisdom of any significant value to take back with us to the Vector Field to be re-absorbed into the consciousness of our Creator. We must be as pure as light itself in order to behold the direct essence of the Absolute. The most blissful moment of existence when we become *One* with the All once again. Christians refer to this part of our Soul journey as the return home to God upon death. This wisdom is of great, significant relevance.

 As individualized units of consciousness housed within earth vessels, we have a significant moral obligation to learn about the functions of Nature. An understanding of the scientific processes of Nature helps us to comprehend human existence, our purpose and our Source. All things that we see are manifestations of energy in motion in a constant state of vibration prompting us to believe in a solid world. We have come to believe that what we see and feel is actually here when really, it is not here. What is here, is our spirit force and our Soul which houses our consciousness and that is what we are. We are Spirit beings caught in numerous cycles of temporary human experience. That is the sum of our biological existence. We are angelic beings trapped inside earth vessels that were crystallized into matter and form through the functions of

Metatrons Cube. No energy is ever dead. Nothing is ever still. All things are in a state of constant vibration and continuous evolution. There is an endless recycling of energy in and out of the physical and cosmic planes. This information is good. This how Nature functions. In addition, there is no *nothingness* because even nothing is something. There is no black, empty darkness in an electric universe filled with electromagnetism, neutrinos and plasma. There is no black, dark veil aside from that of ignorance and lack of knowledge. This wisdom predates the teachings of every currently known religion and spiritual tradition practiced on earth today. It is primordial wisdom carried in the hearts and minds of humanity. This intelligence is encoded within our DNA and our consciousness. Furthermore, this fundamental knowledge is dispersed all throughout the collective mind.

Current wisdom of this great cycle of earths professional wobble as it has been recorded pre-dates 15,000 B.C.E. The research conducted and the vast pool of knowledge compiled by author Freddy Silva, is all good information backed by credible evidence. His academic findings are accurate and his intellectual contributions from the historical metaphysical worlds of Antiquity deserve the highest recognition and honors. Well done. You can check out this excellent resource at www.invisibletemple.com. Freddy Silva is a highly informed researcher and a very enlightening person. Only one word of warning, if you are not interested in being illuminated, enlightened and properly educated about ancient history, ancient tribes and lost civilizations don't even bother picking up his books. They are filled with so much light that it is practically blinding. Intelligent, inquisitive minds will be impressed. Not only is he a best-selling author he is also currently one of the worlds most highly recognized metaphysical speakers.

What we perceive as solid mass is crystallized light vibrating at a very low frequency in the low end of the spectrum of light. The first visible color to the human eyes is red. Red corresponds to the lowest of our 7 main chakra energy centers on our Shakti energy channel. In other words, our spinal column. This base chakra is our root chakra which we can access to ground our energy to earth, center and balance it. Think of this energy as the roots of a tree. The energy associated with our basic animal instincts in effect, respectively finds its home in this energy center. One way to understand the tree of life is to comprehend the alignment of our chakra system and its systematic functions. Our chakra energy centers extend upwards along the tree from the roots to the treetop. Our chakras align from root to crown as they energetically extend upwards along our Shakti spinal column antenna forming the trunk of the energetic tree.

Climbing through the energy centers we experience changes in force, energy and density, as our essence ascends the energy channel in orderly fashion. As we begin to feel more open as a result of working with the energy of each ascending Chakra our levels of human consciousness and cognitive brain functions increase. After our central energy passes through each chakra center finally reaching the crown, our energy is calibrated, aligned, attuned with and open to Spirit. The ultraviolet superconscious causal chakra above the crown chakra can receive energy directly into our consciousness and bio-field. This energy enters our crown chakra as an ultraviolet spectrum of light into our pineal gland. Psychic vision is commonly attributed to this gland and its functions. When our violet crown chakra and ultraviolet superconscious chakra are activated, our brow chakra commonly activates as well. Our brow chakra is indigo. It resonates with Gamma 1 brainwaves and is the proper location of the pituitary gland.

When the pineal and pituitary glands are stimulated in combination, the vesica pisces or 3rd eye within the center of our brain, our cerebral nucleus is activated or opened. Aligned with an activated heart chakra this is our internal gateway to the collective consciousness of the Cosmos. Many refer to this great information bank as the realm of the Akashic records on the astral plane. Once you reach this level of proficiency in your metaphysical energy work you are essentially tapping into the Vector Field. The more knowledge you acquire, the more the spirit forces of the collective mind can teach you. Once you consciously connect to this force through the power of the Inner Self these higher ascended energy forms can show you and teach you a whole lot more about what you already know even just a little bit about. In this way, the information that each seeker of Light receives from the Collective Consciousness will be uniquely tailored and different .The energy of the One mind aligns to the neurological bio-field signal sent out into the quantum field by each individual mind. Our frequency attracts and locks into similar frequencies because it contains information of *like nature.* Once more, Richard Feynman steaks the show with his brilliant theory of *like attracts like.* Now this master physicist is part of the collective mind. I believe that the world is soon to understand what nature and existence truly are and how everything connects. Humanity of this modern age of Light, reason and truth know that this awakening is long overdue. Essentially what we see and experience as solid matter in our biological world including our bodies, plants, animals and the ground we walk on, is simply energy in motion. Light, in various states of vibration. This light energy embodies the forces of electromagnetism plasma, Birkland currents, sympathetic frequency, harmonic resonance, fractals and crystallization. What we need now is for modern science to openly teach humanity all the reasons why we perceive and experience our illusionary hollow world as solid matter. Scientifically we are looking at electric poles, charges and di-poles.

We are dealing with the force of gravity, refracted light, human vision, torsion waves, electromagnetism and so many other factors. Biologists, meteorologists, chemists, cosmologists, astrophysicists, electrical engineers and astronomers could easily form a think tank to create a well researched, evidenced foundation for the absence of solid matter in our universe. It is imperative on many levels that humanity is permitted to discover the hollow nature of our holographic world of cells, atoms, molecules and consciousness. After all, we inherently long for peace and global unity by way of natural human construct. So please go right ahead modern scientific community and show us the quintessential *oneness* of Nature as seen through the brilliant eyes of science. Humanity is ready for this Light. We can handle a fresh new truth.

In book 1 of this 3 volume series, *As Above So Below – Macrocosmic Reflections in a Microcosmic Mirror,* Metatrons Cube was extensively explored. We investigated the functions of the tesseract and discovered that it draws forth the power of electromagnetism from one side and expels the force of gravity and crystallized matter from the opposite side. When we consider the functions of this primordial archetype, we see the Universal Law of Rhythm working at primordial levels of creation. The Law of Rhythm explains the methods of energy in motion. All energy flows in and flows out. Just like the energy generating the holographic universe that flows in and out of the tesseract and vector equilibrium, nucleus of the Cosmic Egg. The essence and basic operational aspects of these divine models of creation were also investigated in volume one of this series. The fundamental aspect of the Law of Rhythm, is that much like a swinging pendulum, everything has a rhythm or sway in the manner of which it moves.

The sea has tides. Everything that goes up must come down. Nations and Kingdoms rise and fall. Rulers and kings live and die, come and go. Generations of humanity are born and then lost in time. Only to be forgotten. The Law of Rhythm teaches us that all things that rise will eventually fall and begin a new. So humanity wishes upon a falling stars. But what do we wish for? This question is key. For our world is what we make it. The energy all around us and within us swings to the left and the right, both high and low. The force of the swing to the left will equate to the force of the right hand Swing. *Newtons Cradle* is a physics perpetual motion machine that was designed to demonstrate the Law of Rhythm. You can find it online in many consumer outlets. What it consists of, is a set of 7 metal spheres attached to cord or wire. One ball is tapped at one end of the 7 spheres and once set into motion through the natural force of inertia, the rhythm produces a balanced swing within the line of metal spheres. Only the two spheres on each end generate the motion while the center spheres remain relatively stationary. *Newtons Cradle* is a device that is commonly known. Most people have seen this device before and will recognize it. As an efficient learning tool it is generally affordable. This fascinating little device facilitates an understanding of the Law of Rhythm by way of experiment through basic observation. In regard to the Law of Rhythm, energy can only diminish so much or in contrast, be built up so high before it must return to its central Natural state. It can only extend so far before there naturally has to be a return. These are the effects caused by the Law of Rhythm. The influence behind the return of energy, is part of the infinite Law of Octaves as it relates to the quintessential architecture of the Flower of Life upon which everything in the physical plane has been constructed.

Extremely low states of vibration only serve to cause a swing of negative energy from one extreme to another. In unhealthy circumstances a swing to the right could be a fall into depression and anxious states while the swing to
the left could only represent an impartial momentary lapse of suffering. This effect is similar to a brief honeymoon stage in an abusive relationship. In the case of finances, a

powerful millionaire can reach the pinnacle of stellar heights in enterprise. Then through his own selfish motives, his own selfabsorbed incentives, through greed, arrogance or materialism he draws forth the power of karmic consequence and his empire plummets into ruin. We see this kind of thing happening all the time in big corporate enterprise, we just don't know the whole inside story. But the millionaire does. And that's all that matters. Essentially, it is none of our business nor do we have a right to interfere with or question the great work of the spirit forces of Nature or the functions Universal Law. The justice behind these operations as it relates to the judgement of God and the Law of Rhythm are beyond our human scope of understating at the highest levels. As aspects of Nature, we are all simply bound by the functions of Natural Law. That relevant detail is what all members of humanity are really obligated or required to acknowledge and respect as a biological species with cognitive rational thought capabilities.

Energy moves in and out of our lives following the primordial code of the Law of Rhythm. This law is in operation all throughout Nature, in the cosmos and in our physical lives at all times. In the case of personal self-healing, transcendence of consciousness and the Law of Rhythm, it is essential that we raise our energy to higher levels and higher vibrational frequencies. Only by doing this can we utilize the Law of Rhythm in a such a way, that it contributes to positive energy in our lives at both extremes of the swing. There are various forms of energy healing tools that have been designed to do exactly that. Binaurals and isochronic audio frequencies listed in my 1st book *Awaken and Ascend* can be very helpful with raising our harmonic frequency. In addition, there are also amazing bio-field technology tools available to us. The final two chapters of this book discuss this bio-tech, what it does, how to use it and where to find it. That being said, assets such as these actually help us develop expanded states of consciousness and awareness. In turn, this state of being contributes to higher vibrations which cause the sympathetic effect of accessing, activating and generating positive energy and higher brainwave frequencies through the Law of Rhythm.

Higher vibrational states center on soulful presence and unconditional loving energies of the spirit forces of Nature found all throughout creation. Achieving a higher state of being is a contribution to an overall healthier sense of self. The resulting effect of this cause is a higher vibratory swing of the pendulum that is governed by the Law of Rhythm. The pendulum still swings only in a more positive energy field. In this way both extremes can generate healthy experiences. Achieving any level of self mastery over mental faculties is a learned, disciplined skill. It requires practice, determination, focus, commitment and a healthy lifestyle emotionally, mentally, spiritually and physically. A compassionate nature and a peaceful sense of spiritual calm usually accompany this mindset and state of existence. It takes a lot more energy to push or operate against the resistance that is endlessly presenting itself in our world. Resistance creates obstacles on the path to our ultimate destination. At times, we may have to push harder against emotional tides to endure a harsh low vibrational swing of the pendulum. When this happens, we can choose to remain stuck in a certain pattern of extremes if we are unable to see a way out of an abusive, unhealthy or negative situation. Being *lost in he mire* in this way can extend for great lengths of linear time. In other words, for weeks, months or even years if the issue remains unacknowledged, unrecognized and unresolved. This defeating condition should not be entertained, encouraged or permitted.

Fortunately, humans have been intelligently designed to eventually tire of the misfortunes and set-backs caused by others or self induced pits of despair. Ultimately, we are left to the task of crawling out of the hole we dug ourselves into, dusting ourselves off and snapping out of the previously permitted negative headspace. The next step is moving forward with a whole new attitude. This initiates a new swing of the pendulum into a better headspace. The ego will always strive for divine rulership over our sense of self, thoughts and outward actions. Give it a crown and scepter and it will indeed take full administrative control over all human functions. Most of the time our ego simply needs a good kick in the backside and a reminder from the Inner Self of who the real boss is. It takes devout self awareness to even recognize the cunning tricks of the ego. Know thyself, this is key to ending all its trickery. It cannot hide from eyes that see. The Inner Self knows all too well when it has an active call to duty. Personal confrontation of our biological ego self requires logic rationality, honesty and humility.

Thank the Absolute for the guidance of our conscience, intuition, moral compass and the inherant compassionate mercy pre-programmed into the human heart. Without this our race would come to a state of ruin and spiritual decline even more rapidly than what is already being witnessed today. Imagine that. Every human being has the ability to break a negative mindset, depressed or anxious emotional undercurrents and raise our energy levels to higher vibratory states. A basic comprehension of the Law of Rhythm and its many operations promotes a desire to accomplish this. Raising our bio-field frequency and elevating our awareness and harmonic energy levels will always alter the negative effects of the pendulums swing. The swinging motion continues in a higher, more positive, harmonic state of vibration. This is how we learn to master the Law of Rhythm and consciously gain management of the energetic swing of the pendulum. If self mastery is your ultimate evolutionary, spiritual goal as it very well should be for each and every one of us, this knowledge is of tremendous significance.

An expanded consciousness generates the higher spiritual states of existence and being that are the ideal condition to strive for. Like all other natural laws, the Law of Rhythm is in operation at all times even when we are unaware of this. The many ups and downs that we experience in life can be attributed to this law and its endless swinging extremes. It is possible to gain some level of control over regulating and shifting these tides of energy that we experience in life. In order to achieve this we must initially establish a balanced mind, body, spirit complex. This step is a pre-requisite of raising our vibrations to a higher level. This accomplishment facilitates control over the effects of the energy generated by the swing of the pendulum. This is one way that we can effectively turn the Law of Rhythm into a valuable asset that we can use to our advantage.

In A World Called Utopia

In a world called Utopia, every Christian is encouraged to study metaphysics and sacred geometry. Regardless of denomination. Because science is essentially and fundamentally the work of God. Quantum physics describes the Biblical teaching found in Genesis. Science explains how the Universe was created in 7 days through the scientific processes of Nature. In a world called Utopia, this knowledge is not taboo. It is understood by all to be the divine work of God. A foundation stone of spiritual knowledge.

In a world Called utopia standard models of science have logical, intelligent competition. They are not set in stone. Instead, these paradigms flex under the weight of evidenced truths and rationality. In a world called Utopia, religious scholars are teaching a scientific origin theory that corresponds to Universal Law. All members of the human race would know that all aspects of creation endlessly repeat themselves in Nature. In a world called Utopia every member of humanity would have intelligent, legitimate cause to accept natural biological patterning upon which creation is fundamentally based. In effect we would logically come to acknowledge the tremendous likely hood that our human race and our earthly abode do not stand alone as constructs of Nature. Humanity would comprehend that it is scientifically impossible for this human race to be the only one of its kind. Indeed, we are not a universal singularity in creation. Such a notion does not exist. In a world called Utopia all members of the human race would celebrate this. All would look to the Stars together as *One* in great awe and magnificent wonder.

In a world called Utopia all would know that *All is One*. Every member of humanity would acknowledge that there are seeds of truth planted within every spiritual and religious tradition known to man that is righteous and good. All faiths stem from energy within the minds of men but that which is rooted in the unconditional love of our Source, is good. Be it Allah, Yahweh, Jehovah, Brahma, God, Source, Creator, the All, the Absolute and so on, all these titles belong to the One and only. Our patron, the one omnipresent Source consciousness God answers to all these names and more. This valid point cannot be stressed enough. In a world called Utopia all members of humanity would realize this.

In a world called Utopia no one would judge another for their spiritual beliefs. All would acknowledge that causes such as non-judgement produce immaculate effects. Indeed, None are to be judged over the other. For no member of this human race has been given the right to do so by the Absolute. In a world called Utopia, all would know that the Source from whence we all emerge and the spirit forces of Nature still govern this aspect of creation. Yet sadly, so many choose to appoint themselves to this vocation of the holy divine, rather unjustly. In a world called Utopia this creed of non-judgement would be accepted as doctrine. All would know that better causes such as amending the weaker aspects of ourselves, will always produce tremendous effects of great value and vast spiritual significance.

In a world called Utopia all members of humanity would recognize the need to regenerate our planet immediately. Gladly, joyously and willingly, each person would choose to do so of their own free will. In a world called Utopia this mission would be conducted with a full respectful awareness that everything we see in Nature all around us is not ours to own but is simply on loan. The aspects of Nature that we utilize for our sustenance are indeed not ours to destroy. Our earth is our Mother. She is a gift to us. In a world called Utopia, humanity would truly recognize and appreciate this. Each member of the human race would respect and revere their privileged role as caretakers of this planet. In a world called Utopia we would gladly be pious porters of the spirit forces of Nature.

In a world called Utopia all members of the human race would respect the fact that our earth is ours to enjoy. It isn't for us to destroy. It nurtures us from birth till death. We are provided with a home. It contains all necessary elements we could possibly ever require for growth and survival in comfort, peacefully, lovingly and quite effectively. In a world called Utopia there would be a tremendous appreciation for this gift. Yet, there is not. Look at the unimaginable amount of damage that we have done to her. Building land mines and nuclear bombs. This is one place we went wrong. Toxic waste in landfill and seas. All of this should bother thee. Damage to the Ozone. Chem-trails in the sky. If we push it any further, our earth will die. From big corporate to lone culprit we are all to blame. All of us deserve this shame. In a world called Utopia, all would know that 2000 years ago, this earth didn't look the same. The population has grown exponentially since then and we have been careless. Toxic air. Toxic water. Toxic soil and a trash heap in the heart of the sea. Animals are in despair. Wild herds are thinning everywhere. Amazon rain forest, cut and burned. In 2000 years what has humanity learned? Sacred rivers run dry and wane. Lands are parched for lack of rain. It's not our planet to ruin. In Gods mind there is no eye for eye. In a world called Utopia, would we save it? Yeah, we'd all try. So, *Order out of Chaos* please. It's up to us humanity. In a world called Utopia, we would stop the insanity.

CHAPTER 2

Determining Sympathetic Vibratory Correspondences Of Frequency & Color Utilizing The Chromatic Scale Of Light

Audible sound is a low vibrational frequency of light. Aristotle, the wise ancient esoteric philosopher was the first to suggest an interdependent relationship between color and light. Color vibrates at a very high inaudible frequency in the terahertz range. When the frequency of color is mathematically reduced through the solar spectrum of light it eventually reaches an audible hertz tone which translates to our human ears as sound. Each color in the solar spectrum vibrates at a different frequency. Our chakra energy centers have various energetic correlations. One of the most commonly known chakra energy aspects is the color assigned to each of the 7 main chakras. Each chakra color has its own energetic frequency and specific rate of vibration. Several systems and charts exist in our academic, metaphysical world that designate specific music notes, chakra colors, brain waves and glands to the human chakras. These findings are devised through various means. The research of Music producer, sound healer, energy worker, Reiki practitioner Dameon Michael Keller offers accurate, modern calculations of these energy correspondences in relation to the vibration of color reduced to sound.

Using the chromatic scale of solar radiation as a foundation, incredibly precise, accurate calculations can be generated through mathematical deduction. This is one of the most reliable methods available for deducing the many energetic correspondences of color, frequency and music notes in relation to our chakras. Using the central frequencies of the audible octaves of notes as a starting point, Dameon Keller set limits to the boundary of each music note as it resonates within its color octave. Utilizing this scientific process he produced a chart of the 12 notes of the chromatic scale based on their color octaves within the Solar Spectrum of light. Dameon has graciously donated the chart below to help you better comprehend the method that he used to reach his scientific results and conclusions. The first column of the chart shows the terahertz frequency of the colors listed by Craig Bohren in his 2006 article "Fundamentals of Atmospheric Radiation". The second column shows terahertz adjustments made by Dameon Keller using a color sound calculator to define the parameters for the remaining shades of color that were not listed with total accuracy. Finally the third column of this chart shows the acceptable range of frequencies for the music notes generated in the audible octave. This chart as well as the further information on the process used to produce his calculations can be found in Dameon Michael Kellers book "Sounds Good, Sounds Great, Sounds Amazing".

Color	Note	THz	THz Adj.	Audible Octave	Central Freq.
(Near) Ultra-Violet	F#	751-800	791-799	724-727	725
interval					9
Violet	F	670-750	782-790	711-723	716
interval					30
Indigo	E		748-781	681-710	686
interval					56
Blue	D#	610-670	648-747	591-680	630
interval					58
Cyan	D	580-610	608-647	553-590	572
interval					17
Seafoam	C#		599-607	545-552	555
interval					38
Green	C	540-580	540-598	491-544	517
interval					34
Lime	B		523-539	476-490	483
interval					14
Yellow	Bb	510-540	510-526	453-475	469
interval					28
Orange	A	480-510	471-498	429-453	441
interval					31
Red	G#	430-480	446-470	392-428	410
interval					32
Deep Red	G		400-430	364-392	378

There are other color-sound correlations systems such as Composer Alexander Scriabin's chart of calculations. In this well known system he places the *C note* at the position of the root chakra and assigns to it, the color red. The results produced in this chart do not resonate with the vibrations or frequencies of the solar spectrum of chromatic light. In addition, these chakra correlations do not follow the natural orderly color of light as it as it emerged from the Vesica Pisces during the 7 primordial days of creation. Essentially, these frequencies do not properly equate to the resultant music note that properly corresponds to our chakras. When using the high terahertz vibrational value of color itself and reducing this frequency by way of mathematical division through the octaves, the music note produced simply does not align with calculations in Scriabin's Color sound chart. Furthermore, this system places a *C note* in the position of the root chakra for one specific reason. In some spiritual traditions the *C note* is taught to be the *root* and the *center*. Here is where the confusion begins. This concept is known to be based on the position of the keys on a piano. On a piano or keyboard the *C note* is both the root note and the center note. In contrast, on the human body *C is the center* and it correlates to our green heart chakra and the planet Venus. The root on the human body is appropriately none other than the *root chakra* which is a red *G# note* that correlates to the planet mars. These are accurate calculations based on logic and knowledge of the proper order in which the 7 primordial colors of light emerged from the Vesica Pisces of the Seed of Life. These findings produced the proper visible solar spectrum color-sound values. This is a result of the precision of the method used. Calculating frequencies utilizing the scientific process of mathematically reducing solar spectrum color frequency values down through the spectrum of visible light, generates accurate results. Dameon Keller has also generously donated a chart of Scriabin's color sound calculations. This chart, along with his work on this subject, can also be found in in his book "Sounds Good, Sounds Great, Sounds Amazing." www.dameonkeller.wixsite.com

Scriabin's color-sound chart
C   Red
C#   Violet
C   Yellow
D#   Glint of steel
E   Pearly Blue & shimmer of moonlight
F   Dark Red
F#   Bright Blue
G   Pearly Orange
G#   Purple
A   Green
A#   Glint of steel
B   Soft Blue

These colors in order create a keyboard as it appears in the following image. You will find the solar spectrum calculations and correlations in the chart that follows. The final chart shows the full range of solar spectrum frequencies as they correspond to the chakras, glands, music notes and brainwaves. These charts can also be found and further explained in Dameon Kellers aforementioned book. Scriabin's color-sound keyboard is pictured first in the following diagrams. Comparing it along side the solar spectrum sound-color chart produced by Dameon Keller shows us an entirely different system than the one produced by Scriabin. In Nature, only one set of colors flows smoothly and correlates properly to rainbows and to the spectrum of refracted colored light produced when you shine white light on a prism.

Scriabin's Color-Sound Keyboard

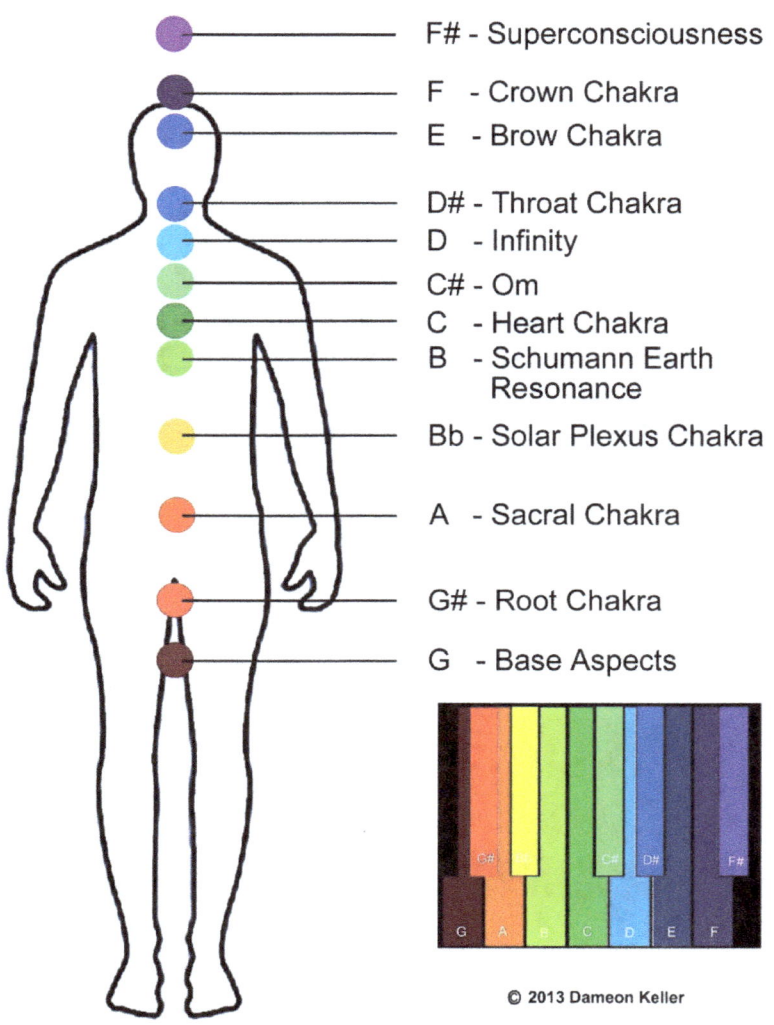

F# - Superconsciousness

F - Crown Chakra

E - Brow Chakra

D# - Throat Chakra

D - Infinity

C# - Om

C - Heart Chakra

B - Schumann Earth
     Resonance

Bb - Solar Plexus Chakra

A - Sacral Chakra

G# - Root Chakra

G - Base Aspects

© 2013 Dameon Keller

Cosmic Keyboard Correlations

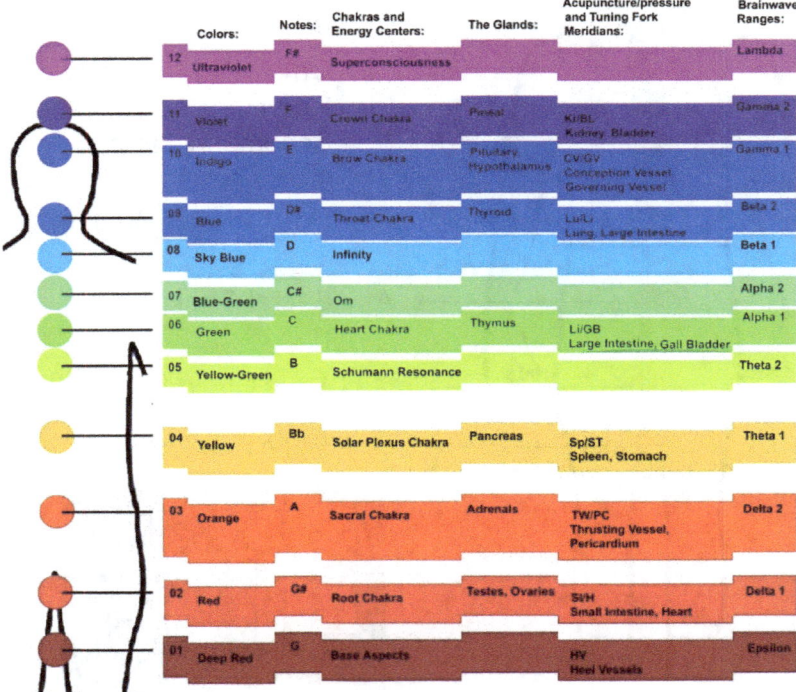

Colors:	Notes:	Chakras and Energy Centers:	The Glands:	Acupuncture/pressure and Tuning Fork Meridians:	Brainwave Ranges:
12 Ultraviolet	F#	Superconsciousness			Lambda
11 Violet	F	Crown Chakra	Pineal	Ki/BL Kidney, Bladder	Gamma 2
10 Indigo	E	Brow Chakra	Pituitary, Hypothalamus	CV/GV Conception Vessel, Governing Vessel	Gamma 1
09 Blue	D#	Throat Chakra	Thyroid	Lu/Li Lung, Large Intestine	Beta 2
08 Sky Blue	D	Infinity			Beta 1
07 Blue-Green	C#	Om			Alpha 2
06 Green	C	Heart Chakra	Thymus	Li/GB Large Intestine, Gall Bladder	Alpha 1
05 Yellow-Green	B	Schumann Resonance			Theta 2
04 Yellow	Bb	Solar Plexus Chakra	Pancreas	Sp/ST Spleen, Stomach	Theta 1
03 Orange	A	Sacral Chakra	Adrenals	TW/PC Thrusting Vessel, Pericardium	Delta 2
02 Red	G#	Root Chakra	Testes, Ovaries	SI/H Small Intestine, Heart	Delta 1
01 Deep Red	G	Base Aspects		HV Heel Vessels	Epsilon

Published & Copyright 2013 Dameon M. Keller

   In the above image, the only change that I would personally make, is to lower the base aspects chakra to a position of below the feet, to properly correspond to the lower causal chakra known to be utilized for Light Body meditation procedures. Other than this, the work proves accurate based on the precise method of the solar spectrum of light calculations. Looking at the work conducted by Dameon Keller, it is clearly obvious that these colors flow together in proper symmetrical order as we see them with our eyes in Nature.

Artists, musicians, philosophers, practitioners of metaphysics, alchemists and various spiritual sects have been trying to bridge the gap between sound, color and frequency for ages. Everything in existence is on the same spectrum of light which generates color and frequencies by moving through various octaves of vibrational energy. Some signals can be perceived by human ears while others cannot. Our eyes and ears perceive sound and light very differently. Both of these energies move in the form of a wave. Sound and light waves can be measured using the Doppler effect model. Light can be in the form of a wave or a particle. There is only one primordial light from whence all things emerge. It moves through the spectrum of octaves and produces the things we see in Nature. This includes all the colors we see, all the sounds hear and the scale of music notes as we know them. All of this, is derived from one divine, quintessential ultraviolet light. This is the profound, intelligent, magnificent method by which the Light of the Absolute functions. Let there be Light indeed.

For centuries, composers, artists, mystics and philosophers and so on, have produced various color-music note charts to calculate frequencies. Multiple correlations have been proposed, studied and experimented on. The various correspondences to color and sound devised, include relations to planets, elements, languages, numerology, seasons, brainwaves, zodiac and more. Intricate sound devices have also been invented over the years. Some of these devices visually and audibly displayed the correlations between color and sound. One of these inventions was a color organ and another was an ocular harpsichord. So clearly, to determine these calculations properly, has been an ongoing project for centuries. This fact alone, demonstrates the tremendous significance that these frequencies and vibrational states of light have on our existence.

Red is scientifically proven to be the first visible color to the human eye. Red vibrates at a frequency just above infrared which is invisible to our vision. We experience and perceive infrared as heat. Red is the lowest vibration, then the spectrum spans orange, yellow, green, blue, indigo, violet and then ultraviolet. This is the natural order in which we see the colors of rainbows and the proper order of the 7 chakras as they are arranged from root to crown. In this accurate model the center is obviously green. The vibration or frequency of color green can be mathematically reduced to an audible *C note*. Red cannot. Red is the root of the color light spectrum, not the middle. The order from whence the colors of light emerge from the Seed of Life is of great significance. This is the quintessential foundational connection that we are looking for. Violet is the highest vibration of the visible spectrum of light. Ultraviolet rays precede violet. Above ultraviolet are blue magnetic XX rays and red XY electric rays. These latter 3 forms of light are invisible to human eyesight. These cosmic rays are thought to be the spirit soul frequency of light. This is not surprising as electromagnetism, microwaves and plasma comprise the vast primordial sea of the Vector Field. In our electric universe the plasma field or quantum field is a highly conductive electromagnetic environment filled with neutrinos and Birkland Currents. Modern scientists keep providing more and more evidence to back electric universe theories and teachings on a regular basis. Eventually such observable proofs will simply become undeniable fact. In regard to color and sound what can be logically deduced, is that visibly accurate results are indeed produced when determining chakra colors based on the solar spectrum of the first 7 primordial colors of light. These calculations also produce accurate correspondences to our glands and brain waves. Imbalanced chakra energy centers cause mental and emotional disharmony and despair. This condition can lead to various states of illness and disease. Aligning these energy centers is a critical aspect of living a healthy existence.

The color of our chakras goes from violet at the crown to red at the root. By calculating the color frequencies of our chakras using the chromatic scale of light, a full range of 12 colors can be determined. Each shade of color resonates with our human energy centers. The most commonly known chakra system is that of the 7 major chakra energy centers. The 7 main chakra energy system is taught and utilized globally by all cultures. All chakra colors do fundamentally derive from the solar spectrum of light, as does everything else in existence on the physical plane. The full chromatic spectrum of frequencies encompasses all 12 shades of color, all 12 musical notes of the chromatic scale and all 12 energy centers in our body. These include the 7 main chakras as well as 5 causal chakras. All of these energetic forces of nature that are housed inside of us, resonate sympathetically within the same orderly system. All of them are fundamentally composed of light.

Precision is the ultimate key in regard to successful energy healing work using color and sound. When color and sound are utilized to attune our chakras to their proper energy levels, accurate frequency and natural harmonic resonance are of vital, critical significance. Color and light have the same symbiotic scientific relationship as sound and pitch. Embracing this detail facilitates our comprehension of the fact that color is in actuality, the tone of light. Audible sound spans at least 15 octaves that are known of in music. On the other hand, visible light spans only one spectrum. Each distinct color emerges as the rate of vibration of the light changes. The light energy moves up and down in the spectrum as it moves through cycles of octaves. All of the distinction between different colors that we see is all coming from the one spectrum. Each color is a frequency of light in motion at varying rates and speeds of vibration which causes each color to come into existence. So ultimately, what we see as color is a bi-product of the original ultraviolet light that initially emerged from the Seed of Life through the vesica pisces. The ultraviolet reduces itself by way of natural fractal deduction producing every single color that our eyes could ever perceive. The reduction occurs in Nature quite naturally as light is transformed into more complex forms of life through various geometric, metaphysical processes of creation. All of this is governed by Natural Law.

Prisms can emit 7 colors of light when exposed to white or ultraviolet light. The light upon entering the prism bends through a called refraction. Disbursement of light at different angles effectively generates all 7 colors in proper harmonic order. By comparison in the primordial model of the Vesica Pisces, ultraviolet light streams forth from the plane of the Absolute. Next, all 7 colors are generated during the construction of the Seed of Life. As you can see, the prism is a tool by which to investigate this profound optical effect which essentially corresponds quite nicely to the architectural components of the Cosmic Egg. With our human vision we can see both light and dark shades of a certain color. The variation of shades is due solely to the distinct, different rates of speed at which each color vibrates. As the frequency of light changes an entire spectral range for each color is generated. For example, the color blue and the many shades of blue that we see are all comprised of frequencies in various states of vibration which ultimately determine how we optically view the color blue. Our biological world is filled with a great variety of beautiful colors that come in a vast variety of shades. Only one spectrum of light is responsible for creating this magnificent artistic palette and array of stunning colors. From this vantage point all things in our universe and in Nature all around us, which do indeed emerge from the same source light can truly be seen as One. Everything in creation is connected and is clearly of very intelligent design.

Within the solar spectrum, the vibrational rate of violet is slightly double that of red. The range of colors we optically perceive is dependent on the rate of vibration of light within its octave range. Sound waves differ from light waves though fundamentally both forces are composed of the same energy. As previously noted, sound and light are measured in hertz and color is measured in terahertz. The standard range of audible human hearing is 20hz-2000hz. Thus we can see color but we cannot hear it. Colors in our atmosphere have a tremendous effect on our health, on our state of mind and overall sense of being. Colored light is used as an energy healing tool in Chromatherapy techniques. The frequency of light utilized demands total accuracy if one wishes to generate specific results. Miscalculated frequencies will not alleviate or improve any biological conditions. In Chromatherapy healing practices a chakra attunement goes from red at the root to violet at the crown.

Various isochronic and binaural brainwave entrainment sessions available on the global market are generated using the major sol-fa scale of 7 notes. This scale starts with the C note as the root note. This system of frequencies is based on C as the root and center note on a piano keyboard. That being said, a keyboard is an inanimate object and the human body is a biological organism with entirely different harmonic frequencies generated by light. I suppose at this juncture one could legitimately argue the fact that inanimate or not, the piano itself is essentially composed of light as well. However, we would ultimately be getting a little off track somewhat in our rational thinking process. In great contrast to a piano, the human body is a living being. In reference to our biological body, the heart is the center and the root is the root. This information is essential foundational knowledge for energy workers and practitioners alike. In Reiki healing, Chromatherapy and any other form of energy healing using light or sound energy, the vital wisdom of proper harmonic resonance is key.

Scriabin's color chart was not the first example of a system that correlated Red to a *C note* and our root chakra. Many other charts have been designed that incorporate this calculation. Again these systems are based on a keyboard whereas on our human body the center is the heart, the 4th major chakra in line from the crown. The 4th color to emerge from the Seed of Life is green. In proper order the 7 main colors are violet, indigo, blue, green, yellow, orange and red. Based on accurate calculations of the vibrational rates of light that comprise the solar spectrum, Scriabin's color-sound calculations do not correlate. His findings do not necessarily attune to human sympathetic resonance and our proper bio-field harmonics. Since all things in nature and creation are quintessentially connected through the solar spectrum of light, these findings pose a problem in regard to accuracy. The C note resonates with our green heart chakra as well as with our Alpha 2 brainwaves. C 256hz is in the 3rd octave of the scale. Earth is in the 3rd harmonic in its orbital position as 3rd rock from the Sun. Precise color, sound, chakra correlation generated using the solar spectrum of light produces C as green and G# as the red root. Acknowledging this fundamental detail facilitates production of successful results in regard to energy healing techniques utilizing the modalities of sound and light. Balancing procedures such as those used in Chromatherapy and Reiki healing can only produce the expected beneficial results by incorporating accurate harmonic frequencies into the technique.

Ultimately, music notes are frequencies derived from the direct relationship between sound and color which are both energetic by-products of light. To find the audible tone of color the mathematical procedure is to divide the terahertz frequency of color down through the octaves by two. After much division we eventually reach the audible sound of the colorin the hertz range. To this frequency we assign a corresponding music note. This form of Chroma-acoustics produces the most accurate calculations of the color to music note relationship. Only precise accurate frequencies will entrain our brainwaves which resonate within very specific pre-set frequency ranges. Positive successful results in any form of energy healing can only be rendered utilizing precise frequencies. The various energy healing methods and techniques that incorporate light or sound can deliver the most effective healing when properly administered under the right conditions. Only specific harmonic frequencies will activate our chakra energy centers which in turn engage the proper corresponding brainwave state. Together these two forms of healing contribute to spiritual evolution, expanded awareness and transformation. This information is of vital significant value in regard to achieving desired results using bio-field technology, isochronics, binaurals or Chromatherapy healing methods. When calculated and used properly, harmonic resonance can produce immaculate, transformational results. In great contrast, when improper frequencies are utilized results can range from catastrophic to ineffective. In this way the resulting effect can cause more damage than that which existed in the first place.

In order to verify the accuracy of his color-sound solar spectrum correlations mentioned earlier on in this chapter, Music Producer, Sound Therapist, Dameon Keller conducted a tremendous amount of research. He explored various scientific resources related to color and frequency, painstakingly calculated the vibrational rates of the frequencies of color and meticulously evidenced his theory. The resulting chart that he generated shows the color-light music note correspondences of the chromatic spectrum. The results he produced appear to be accurate as they do generate frequencies that properly align to our known brainwave frequency ranges. Ultimately this scientific method is the most precise system that we have of defining harmonic frequencies that sympathetically resonate with the human biological organism. Using the solar spectrum to calculate the scale of audible music notes gives you all notes in proper order as they correlate to light. There is no method on earth that will produce the same key values. Thus, it is an accurate system to follow. It is the one I personally use with good results. The 12 notes arranged in order as they derive from Nature creates the chromatic scale which does not generate a flowing scale. When using the solar spectrum of primordial color as a guide, the set of chakra, music note, color correlations that is generated corresponds to our brainwaves and glandular system and is extremely accurate in that way.

When mapping the 7 main chromatic chakra frequencies to the human body in their respective positions, the remaining 5 notes of the scale fall right into place quite naturally. These represent our energy centers which are known as causal chakras. Together they form the whole chromatic scale as it corresponds to all 12 of our chakra energy centers. This work can be seen in the full body chakra chart at the beginning of this chapter which is the work of music producer, sound therapist, Reiki energy healer, Dameon Keller. In the following image we see the seated body showing the 7 main chakras.

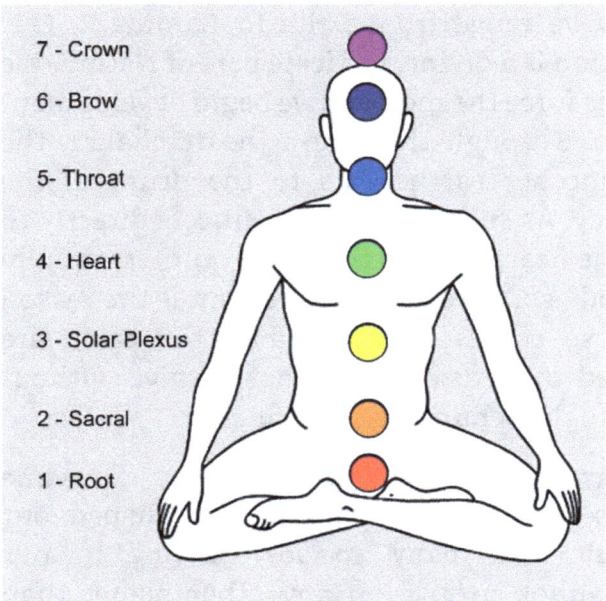

7 - Crown

6 - Brow

5- Throat

4 - Heart

3 - Solar Plexus

2 - Sacral

1 - Root

In the chromatic spectrum the 7 main chakras naturally align to the 7 base colors of light as they emerge from the vesica pisces. The frequencies that align with our chakras fall into a very specific order. Moving from root to crown the 12 chakras begin at the bottom with 3 half steps at G#, A, Bb. These notes as such, resonate with our 3 lower chakras. Red equates to the root, orange to the Sacral Chakra, Yellow for the Solar Plexus. These are followed by the note B which is normally not included in the popular well known 7 chakra system. The B note refers to a causal chakra. This is an energy center above the solar plexus and below our heart chakra. The color of this chakra is a bright, vibrant yellow-green that resonates with the Schumann frequency of the earth. This chakra is a causal heart. It is located within the energy field of our main heart chakra and our biological heartbeat. Our causal heart is connected to the Scientific forces of our planet at all times during our biological existence. The Schumann Resonance is 7.83Hz. Throughout the day Schumann moves through 8 peak frequencies. These frequencies match and align with our human brainwave range from Delta to Gamma 2. Our biological composition is a divine, intricate part of Nature. We are connected to this force the moment we begin development as a fetus in utero. Through this causal heart chakra, the heartbeat of our mother earth beats to the drum of the Schumann frequency. As such, our heart connects directly to hers in this way. The heartbeat of our planet quintessentially beats within our bodies. This sacred attunement is the reason that humans, whose bodies are formed of the earth are energetically aligned to Nature and cannot survive without constant exposure to the Schumann resonance.

Our biological earth vessels are composed of the same elements, minerals and energetic forces that shaped our planet. As a result, the many species existing in our animal kingdom cannot survive on any other planet than this earth, which they harmonically resonate and attune to. Furthermore, our specific species of humanity cannot survive on any other planet in the milky way galaxy except our earth. We are bound and anchored to our planet by a locked in frequency. As are all other forms of biological life on earth including our plant kingdom. This critical factor is the reason why all space shuttles must be equipped with a Schumann Resonance Simulator if the astronauts are to survive their journey outside of earths atmosphere. So until technology reaches such an advanced level of sophistication as to invent the Schumann back pack, our human race isn't going very far. By way of Natural Law, this information would also apply to any other human species in the universe who exists on any other earth like planet, in a model solar system just like ours. Can we rationally, logically and intelligently assume that our solar system, our race and our earth are universal models? Yes, and with all due reason. Especially when one takes into consideration, that all manifested things come into existence as a result of quintessential patterns of creation found in Nature. The Law of Correspondence is the key.

By the automatic primordial standards of the Law of Correspondence, all things in creation endlessly repeat themselves. Nothing is exempt. Galaxies, solar systems, planets, stars, human and animal species. What do these other earths, human races and different types of animals look like? Do they look like aliens, giants or people with long legs like a giraffe? Are they very similar to us? It's a profound concept to contemplate and a very exciting one as well. The truth is, we don't know what they look like. We haven't met anyone else yet other than earthlings. Modern Science may someday provide a means to do so but for now, we have the keen sharp eyes of the James Webb Space Telescope scanning the cosmos for other forms of life. So yes, the potential that our solar system is a model in a physical plane where creation is an infinite repetitive process of reflection from macrocosmic to microcosmic levels of manifestation, is highly probable. From a metaphysical perspective, it's a logical conclusion to draw. Are we alone in the Universe? From the vantage point of the Law of Correspondence as all things are governed by Natural Law, that would be scientifically impossible. When one considers the wondrous nature of the endless creative potential found within the vector field, the possibilities of life existing all over the universe are unlimited. The axiom, *As Above So Below* is always teaching us that by way of the Law of Correspondence, creation is one big cycle of various scientific processes of Nature locked into infinite states of repetition. Everything in Nature is connected. All things in creation are primarily and fundamentally *One life Force* comprised of primordial light vibrating at different speeds. In addition, all things in creation that are manifested of light contain the spirit force and consciousness of our creator. We are immortally connected to the mind of God by way of intelligent design and divine construct.

All of the magnificent work of our Creator is governed by

the omnipresent forces of Universal Law. Our biological earth vessels that house our sentient self are of earth, from earth and perfectly aligned with Nature. As a result of this divine biological connection, our harmonic resonance and bio-field energy are sympathetically attuned to that of our planet. As such, when we are sick as a race fighting or going to war, being cruel, abusive or hostile towards one another, not only is the harmonic vibration of the human race reduced to lower sympathetic frequencies but the vibration of the planet is lowered along with it. Case in point, if the biological life forms on the surface of the planet are sick, the planet feels this illness in response. Cause and Effect is one law that often plays a huge role in the process. In contrast, when we thrive as a race living in accordance with Natural Lawwhile being respectful porters and caretakers of our earthly abode, both the harmonic frequency of our species and our planet are raised in effect. When the earth is in its most prime optimum condition of health and wellness this energy extends to all biological lifeforms that live within its environment. In this light, it is obvious that humanity makes the world what it is. Our global environment is currently in a state of rapid decline. This shows us just how much we care about our human responsibility to protect our only home from the destructive habits of our race. In truth, we have much restoration to conduct on so many levels. Environmentally, physically, spiritually, emotionally, scientifically and even in the case of religion there is great need of repair and reformation. The planet desperately requires medical aide. If we wish to leave behind a healthy home for the youth of our nations, the next generation, there is much work to be done. In this significant revelation we find countless reasons to save our earth from the immense amount of damage that we are responsible for. Even our animal kingdom suffers and diminishes by the day.

The same scientific universal principles and models of creation apply to everything that exists in the diverse pool of biological lifeforms that are dispersed all throughout Nature. All living organisms on earth are attuned to the Schumann resonance and biologically aligned to the earth. This sympathetic connection exists everywhere in creation. After we reach the Schumann causal heart in our chakra energy column, we arrive at the color green in the solar spectrum and our divine center. The domain of the green heart chakra resonates with a *C note* and the planet Venus. Here at the spiritual center, is exactly where the heart belongs. This sacred center is not red. The color red has no business at all showing up in the solar spectrum of light in this general location. Red is the 1st visible color of the visible spectrum, not the center. In the solar spectrum red resonates with a G# note, not a C note. In this order, the heart is the center and the root is the root. This knowledge is very significant. Directly above our C note heart chakra, we find a blue- green energy center. This is our AUM chakra which resonates with a C# note in the solar spectrum. This chakra is also right in line with our main heart chakra. As a result this energy center resonates within the same energy field as our 2 other heart related chakras. It constitutes the essence third heart center and second causal heart. Combined, these three chakra centers energetically contribute to one force which can be viewed as our tri-heart chakra complex respectively.

Our C# AUM chakra connects to universal oneness, radiates unconditional love, peace, unity and corresponds to compassion. This chakra resonates with alpha 2 brainwaves. Our main heart chakra center resonates with alpha 1 brainwaves and our Schumann causal heart resonates with theta brainwaves. Together, these 3 chakra energy centers are a very powerful force to activate in unison. This heart center is the domain and temple of our Inner Self. You can cross the threshold of this temple, but doing so requires an attunement and perfect alignment to the Inner Self. After all, that is whose door we are essentially knocking on. Generating this alignment is initially facilitated through the use of Alpha 1&2 binaural brainwave entrainment audio tracks. In addition, these effects can also be produced through regular utilization of the bio-field technology program reviewed In the 2nd final chapter of this book. Our spiritual alpha brain hemisphere is essentially our creative subjective mind and the mind of our Inner Self. Attuning to our heart energy center on a regular basis promotes a general state of existence that resonates with Nature. This connection essentially generates higher harmonic vibrations, expanded awareness, compassion and of course, a deeper sense of unconditional love aligned to our Source. All of these three chakra centers exist within the sphere of energy encompassing our physical heart organ. While our physical, biological heart does not resemble the literal commonly known heart shape, it is indeed associated with unconditional love of the spiritual alpha mind. Ultimately, our physical heart takes on the shape of an ovoid. The divine structure of our heart organ is a similar architecture to that of the Cosmic egg. The heart, is the main biological engine and nucleus of our earth vessel. The nucleus of the cosmic egg is the main engine that generates the holographic universe we live in.

Next in line after our AUM heart chakra, we reach what Dameon Keller has termed our infinity chakra. This chakra energy center facilitates both inward and outward communication with biological and spiritual life. This light blue colored infinity chakra encompasses the energy of the infinite processes of the spirit forces of Nature as well as the fractal and sacred archetypal energies that govern all creation. This causal energy center resonates with a D note and is attuned to our vast, expansive oceans and our limitless sky. After this, the scale follows another three half steps which comprise the three final notes of the chromatic scale that relate to our seven major chakras. Next in the line up is our royal blue throat chakra. This energy center correlates to a D#

note. Following this, we reach our Indigo brow chakra which is commonly referred to as our third eye. This energy center resonates with an E note. Next we arrive at our violet crown chakra which corresponds to an F note. Finally, we reach our final chakra at the end of the chromatic scale of light which is a causal chakra related to superconscious energy. This chakra energy center resonates with an F# note and ultraviolet light. Our superconscious chakra is one of the main chakras associated with Light Body activation practices and communication with the collective mind.

The results generated in this 12 note chromatic scale along with all sympathetic correspondences are based on chroma-acoustic science and harmonic vibration. Color, tones and music notes can all be determined through simple mathematical calculations of light frequency. All energy in motion moves through one solar spectrum of light at varying rates and speeds of vibration. The 12 note chromatic scale produced by Dameon Keller combines data from multiple credible sources. Quantum biology, metaphysics, ancient and modern philosophy, various spiritual and scientific theologies, rationality and logic. His calculations were generated utilizing a sound scientific mathematical procedure. Only very specific frequencies will sympathetically resonate with human brainwaves, the human bio-field and our chakra energy centers. In regard to sound therapy techniques, scientific precision is fundamental. Utilization of improper harmonics will never produce harmonic states of being. In any case, accurate harmonics will always generate expected, desired results. There are specific frequencies that resonate with our whole body. These sympathetic harmonics can alter our states of mind and being with profound positive effects. Dissonant, inharmonious frequencies will only alter our states of mind and health in a negative manner.

Brainwaves are frequencies that link very intricately to

their correlative music note, rate of vibration and color of light within the spectrum. In Chromatherapy healing, results cannot be achieved unless the correct frequency of color corresponds to the chakra energy center. Utilized properly, this form of energy healing has the ability to aid in cleansing our chemical system, biological body, cells and resonant bio-field from unwanted, unhealthy energies. It is imperative to acknowledge that there is only one set of correct frequencies, harmonics, colors of light and music notes that match our resonant energy field. This signature is found in the brain. Each brainwave state will always correlate to only one set of frequency ranges. They cannot be properly activated in a healthy positive manner through exposure to any unnatural, random music notes or frequencies. Neurological science has already determined that each brainwave state responds to one specific music note, one frequency range and one color of light. The solar spectrum of primordial light from whence all things in creation emerge is the most accurate asset and tool that modern science and practitioners of metaphysics have access to.

The most common cause of ineffective energy healing results, is often contributed to the confusing, common color-chakra correlation error of positioning the C note as the red root. Unlike an inanimate piano keyboard, color and frequency are applied to the human body in a far more intricate fashion. The theory and model presented herein, designed by Dameon Keller generates accurate results utilizing a logical, rational, scientific approach that is provable. It can be tested, experimented upon, observed and verified by other academic professionals in various fields of scientific study. He simply applied a logical modern approach to the physics and metaphysical properties of solar spectrum light. In the end, the colors orderly align with the chakras as one energy system in symmetrical order which makes perfectly rational, logical sense. Furthermore, this system does generate proper harmonic vibratory correlations to our commonly known, accepted brainwave ranges. In Dameons own energy healing practices as well as in those of his solar spectrum sound therapy students, this model has been proven effective producing desired results. It acknowledges all 12 energy centers in the human body, all 12 notes of the chromatic and semi-tonal scales, all 12 initial shades of color produced by the solar spectrum, all in correct order and in proper harmonic resonance with our full range of 12 brainwave frequencies. All of these intelligently designed, sympathetic forces automatically sync into alignment quite naturally in this system, just as they automatically do all throughout the biological world of Nature. All around us and within us these frequencies vibrate in a state of perfect harmony by way of divine orderly, conscious design. Everything in existence is fundamentally comprised of light in various states of vibration. All of this is governed by Universal Law. Bridging the gaps by connecting the harmonic dots between these invisible energy fields of like nature is key wisdom.

Electromagnetism is a force comprised of magnetism and electricity. This force is naturally inherent within the construct of every biological creation on the physical plane. Electromagnetism emanates outwards from the divine center of any biological organism. The field created by this energy flow is scientifically described as a resonant bio-field. This energy moves outward away from the life form emitting it. It moves in the form of concentric rings. These rings flow out in perfect order alternating red xy electric rays evenly with blue xx magnetic rays. Human bio-field energy is comprised of these perfectly ordered, alternating red and blue rings which will extend up to 5 or 6 feet away from our body into the space around us. This energy field is both capable of receiving and projecting information. The resonant bio-field forms the shape of the well known symbol of a common circular bull's eye target or the ring pattern of a dart board. The multi-concentric ring pattern is an energy form model that is repeated all throughout Nature and recorded by every age of man. This is evident in the shape of a solar system, structure of an atom or a ripple in a pond formed by a stone dropped in the water. The multi-ringed concentric circle was also very common symbolism used in the construction of ancient ruins and stone circles which can be found all over the earth. Additionally this symbolism was the model for the blueprint of the popular well known capitol city of Atlantis, Poseidonia. This concentric ring structure is also a common symbol associated with earth quakes and seismic activity.

Bio-field energy is dispersed all throughout the cosmos as well as within the diverse pool of biological creations found here on earth. Concentric ring pattern bio-field signatures are uniquely coded for every type of life form in existence. The core of this code becomes the blueprint for further versions of the same type of life form to reproduce on the physical plane as an active part of Nature. Each new emergent species or type of life form will have a slight varying degree of vibrational frequency. This distinction in rates of vibration is what gives each creation the unique characteristics that cause all manifestations to differ from one another. This cause results in the effect of the illusion of individuality, separation and division. Where else do we see rings in the cosmos? We see them forming around highly conductive brown dwarf stars, like the planet Saturn. Ultimately bio-field energy rings form around everything that has an electromagnetic torus. There are other planets, celestial luminous objects and stars in the universe that demonstrate electromagnetic ring like structure as well. In chapters 11 of this book we will further examine the Nature of sympathetic bio-field resonance. Exploring the basics of how it functions helps us learn how to tap into these powerful invisible energy fields that we interact with on a daily basis. But first, let's journey into outer space to investigate, study and explore the macrocosmic world of frequency and vibration. Welcome to the Cosmos. A universe filled with quasars, massive galaxies, clusters and galactic plasma jet stream ejections governed by the law of correspondence, awaits us.

Are We Alone In The Universe?

The light is out there. When the human race discovers the truth about the origin of our existence as it metaphysically corresponds to Natural Law, it begins. A global spiritual awakening will be initiated. Humanity can then establish the logical knowledge that we are not alone. Not in a universe where all processes of creation endlessly repeat themselves. This is occurring in an infinite cycle all throughout creation from macrocosmic to microcosmic levels of existence. The conditions under which we *could be* the only human race in the universe simply do not exist. All of creation is structured and built upon quintessential primordial models and divine scientific systems of manifestation. These systems repeat themselves all throughout the diverse world of Nature. Creation is dual and reflective by construct. It reproduces itself creating various lifeforms through the same patterns and scientific processes over and over again.

Bearing this in mind, can there be only one version of any individual form of life biological existence? No, logically there cannot. Within creation we find an abundance of lifeforms that show observable likeness to other biological life forms found in Nature. We can literally see these patterns all around us. Ultimately it is logical, rational and safe to deduce that the human race itself, represents the same divine constitution found within the same paradigm and the same reality of existence. Our planet equates to a spec of dust in comparison to the massive size of our milky way galaxy. Our galaxy is only one of the numerous galaxies that our cosmos contains. There are thousands of galaxies in the massive, expansive realm of space. Our entire human civilization can be seen as a miniscule piece of micro-dust in a cosmos teeming with dust and matter that has potential to become biological life. It would be egotistical, arrogant and unintelligent for us to think that the massive, expansive universe we live in was built solely to house our tiny little race and our tiny little earth. For us to be the only human race in a universe of infinite possibilities outright defies the omnipresent functions of Natural Law. Essentially, if we were the only human race in existence, this would also defy the laws of quantum physics. This fact should be well known by all. In light of this knowledge, the odds that we have human family members in other parts of the universe even inside of our own galaxy is a logical, highly intelligent rationalization. To find the truth, look to the stars. With the aid of the James Webb Space Telescope humanity can potentially peer into other worlds. Are they much like ours? What do these other biological beings look like? Do they look like us? To acknowledge, believe in and abide by Natural Law is to live in awareness of truths such as this. Will JWST find life out there in The Cosmos? We will just have to wait and see. So many magnificent, spiritually enlightening discoveries await humanity.

CHAPTER 3

Physicist Wal Thornhills James Webb Space Telescope Insights, Predictions And Revelations

The James Webb Space Telescope has stimulated the world of scientists, cosmologists, plasma physicists, astrophysicists and professionals from various academic fields. Since the 2022 launching of JWST scientists have been quite busy meticulously researching images sent home to earth from deep space. Amongst these researchers was physicist Wal Thornhill. One of the world's most inquisitive, highly intelligent philosophers and scientific abstract thinkers. His observations yielded further provable evidence of electric universe theories which he shared in a 3 part video series review of JWSTs discoveries. Regretfully for all who loved him, Wal passed away at the age of 82. His 3 part series aired in 2023 and can be found on www.thunderbolts.info. Upon examining the images of deep space, a scientific research team at the university of Missouri apparently found a galaxy with a redshift of 20. Why did Wal Thornhill find this discovery to be so phenomenal? Using the standard model of redshift this indicates that the galaxy would have formed 180 million years after the big bang occurred. This posed a problem as this conclusion defied commonly known scientific teachings regarding galaxy evolution. It is far earlier than the origin of galaxy formation is said to be. This enigma has some scientists perplexed and questioning known standard models of early galaxy and star evolution. Thornhill believed that physics has been made so overly complicated that its quintessential simplicity may have been overlooked. Within the mainstream scientific community there is even a passive, quiet questioning of the standard model of dark matter.

Wal pointed this detail out in one of his JWST video reviews by quoting a recent scientific paper entitled, *Cosmology Intertwined A Review of Particle Physics Astrophysics and Cosmology, Associated with the Cosmological Tensions and Anomalies.* This research paper was authored by 150 academics from various fields of the professional scientific community. It discussed inconsistencies and possible cracks in standard cosmological science. Ultimately, this document clearly indicates that respected modern scientists are slowly beginning to express a need for new physics that better explains reality beyond the limited scope of the accepted, modern scientific paradigms. Wal Thornhill continuously expressed the fact that math is not physics. Mathematics can only describe the results of physics. The teachings of modern physics are filled with endless complicated mathematical equations. As a result the subject of physics itself has become so difficult to comprehend that most people don't even have an interest in studying it. A great need now exists for the science to be simplified.

What I do know for certain, is that one mile long mathematical equations can indeed describe the results of physics. Regardless of this, intricate, complicated math must take second place to sacred geometry, platonic solids, electromagnetism, spheres, angles, straight lines, simple fractal equations, cymatics, frequencies and so on. The math involved in reducing the color of light to audible frequency is simple division. Basic multiplication is used to calculate C-Scientific pitch and the various stages of embryonic cell division. To calculate Fibonacci we use simple addition. The numbers are there, but the math itself is easy enough that anyone can learn it. Forces such as these compose all matter that is governed by Universal Law. Origin theories and physics that align to Natural Law describes our purpose, our grand design and our existence.

The aforementioned paper goes into great lengths discussing various scientific factors that remain outstanding regarding standard model cosmology and astro-physics. Many issues in these areas which still remain unsolved are in need of further research and await verification. Many scientists agree in the imminent alteration and reconstruction of standard scientific models such as cold dark matter and the cosmic microwave background theory. Inconsistencies in standard model paradigms may be causing the need to re-define certain core scientific beliefs. This is a necessary requirement if modern science wishes to meet the fast changing pace of our modern technological world in an age of reason and light. Humanity needs answers that work. Concepts that align with accepted, quintessential scientific knowledge of our metaphysical electric universe will suffice.

Famous plasma cosmologist Anthony Peratt has recently stated his opinion that a growing number of observations are now questioning big bang cosmology. You can find his thoughts on this subject in his paper entitled *Plasma Cosmology Part 1*. Plasma physics focuses on the known presence and observable functions of electromagnetic plasma in our universe. Electric universe teachings and philosophies define the great blueprint of creation. From tiniest sub-atomic particles in existence to the largest of biological systems, all things in creation are electrical by way of standard construct. Not only is science on the brink of a revolution, in many ways, it may have already began. Many of the world's most sharpest and brilliant scientific minds are slowly and gently beginning to question some of the inconsistencies of known standard models of science. As such, modern science may be on the verge of an inevitable long awaited paradigm shift. At the very least, the need for this refreshing change is logically being recognized in the professional scientific arena. The science of the electric universe incorporates geology, biology, chemistry, atmospheric physics, quantum physics, plasma physics, electrical and bi-molecular engineering and so on. The current existence of standard models should not limit humanity from expanding our awareness to explore other rational ideas, logical theories or intelligent concepts.

In any case, an outright opposition to standard models will never get anyone in the academic community praised or raised to intellectual heights. At all costs we are ultimately dealing with systems set in stone be they efficient, questionable or not. Developing an understanding of creation through an acknowledgement of the holographic universe theory incorporates a study of sacred geometry. The architecture of the Isotropic Vector Matrix is constructed by the processes of Universal Law. The primary function of Metatrons Cube is the crystallization of light and electromagnetism into biological matter. In order for humanity to make rational sense of our purpose, existence and our Source, it is imperative to acquire knowledge of the metaphysical science of Creation. From the construction of the Seed of Life right down to the structure of a biological pinecone, we find infinite fractal patterns all throughout creation. This fundamental wisdom is inherently encoded within our consciousness, DNA, bio-field signature and cerebral cells. In Many ways, to acquire this knowledge all one really has to do, is remember. There are many ways to trigger theses memories that are locked within our biological organism and sentient spirit. One way is through the use of frequencies such as binaurals, isochronics and audio bio-technology.

When it comes to acquiring knowledge in an objective manner

all logical, intelligent views are worthy of consideration. No logical, intelligent intel is bad data until proven to be incorrect. Theoretical Science is based on hypothesis, concepts, ideas, models, research, study, evidence, ancient knowledge, philosophy and so on. It is comprised of any intelligent theory about nature or creation that is waiting to be tested, researched, experimented upon and proven either correct or incorrect. If the potential conditions exist by which to do so, it becomes the main obligation of academic science to see it done. It's that simple. In regard to building a credible belief system, each of us must look at the scientific evidence, facts, logic, reason and intelligence behind any concepts that are relevant us. Humanity is equipped with freedom of speech, self expression, free will and intelligent, rational cognitive thought abilities. Inherent within the divine center of our biological earth vessel is a heart and conscience that know the difference between love and hate. Each of us have a built in moral compass that can distinguish between right and wrong. There are no excuses for our ignorance at this stage in the game aside from lack of proper, accurate wisdom. All humanity truly requires in these changing times is truth, light, hope and mercy. United as one in radiant light attuned to Natural Law, our species would evolve to whole new level of harmonic vibration. In the face of wisdom and truth all members of the human race would have a reason to look up to the stars together in wonder and awe. To behold and marvel over the work of our source is to embrace the knowledge that we are not alone. If everyone knew this would we all look on in unity with hope, excitement and great wonder? I'd like to think so. For therein lies the power of the light of truth. The need for faith is desperate. The vote for truth is unanimous.

Wallace William Thornhill
1942–2023

A brilliant man, Wal Thornhill loved science and saw far beyond the veil of ignorance into the luminous stars of all galaxies everywhere. We were gifted with his presence. Enlightened by his wisdom. Radiated by his light. For those of you who did not know him, the opportunity still exists. For indeed he has made his mark on this generation of mankind. His knowledge has been left behind for us to further explore. For science to further research, experiment upon, study and prove or disprove. Fortunately the same advantage applies to the life works of the great scientific explorer, Immanuel Velikovsky whom Wal Thornhill greatly admired. In the great beyond of the cosmic plane are these two luminous minds watching JWST in action? It's very likely that they are. Much love to you from earth, Mr. Wal Thornhill. Information regarding Wal Thornhills research and JWST predictions can be further researched on www.thunderbolts.info.

CHAPTER 4

The Dawn Of A New Age Of Light. Carefully Questioning Scientific Reality As We Know It.

Prior to the launch of the James Webb Space Telescope and before he passed away of natural causes in 2023, Wal Thornhill made three Thunderbolts videos in which he made several predictions as to what the James Webb Space Telescope would reveal. In addition, Wal highlighted the significance of this highly advanced piece of technology to further validate and evidently support electric universe theories and teachings. The videos were presented on the Thunderbolts website www.thunderbolts.info, narrated by Stuart Talbott and released on June 19, 2021, September 5, 2021, and January 1, 2022. In this chapter we will review and explore his predictions in my great cosmic search for scientific models of creation that align to the metaphysical Universal Laws of Nature. Bearing in mind that an objective approach to science is of great significant value, we will leave no stones unturned in our search for truth. We will not judge information as bad input until proven to be such. We will not discriminate against sources in our search for Light even when facing radical outside the box concepts. You never know which fresh ideas, old revelations, new angle approaches or scientific epiphanies may serve to aid in ushering an age of reason to the forefront.

Wal Thornhill was fully aware of JWST's ability to unravel some of the mysteries of the universe and evolve modern science to the next level. Equipped with highly advanced 24 karat gold imaging chips, infrared technology and various other forms of high tech viewing tools, JWST is taking immaculate images of the cosmos for humanity to behold. Wal predicted that there would be much talk about redshift in the world of science to follow the release of these images. For professional, student and hobbyist astronomers alike there is excitement in the air. Redshift values and cosmic phenomenon will be discussed throughout the next three chapters. That being said, let us first establish a basic, general understanding of the scientific nature of redshift. Quasi stellar objects contain high amounts of compound chemical gases. These elements and chemicals such as hydrogen for example give off signature types of light. When an object is measured through waves of light a shift in frequency occurs based on the chemical composition of the cosmic object being measured. The resulting spectral emissions or absorption lines of the wave lengths of light emitted, will correspond to light given off by elements or compounds when compared in a laboratory. In regard to redshift this refers to an increase or spectral shift of light waves.

Astrophysicists have three different ways of measuring redshift. The first method of measuring redshift has been derived from the model of an expanding universe. It is based on the velocity of an object relative to the rate of speed at which space is said to be expanding. This model indicates that high luminous objects are located at great distances away from the observer. Lower redshift would thus indicate celestial objects that are much closer to us here on earth. The second way that modern scientists measure redshift is by utilizing the Doppler effect. This process encompasses a measurement of the objects motion in comparison to the motion of other objects in the cosmos. The Doppler effect measures distance of an object from earth using light waves. In this method, red is factor of vast distances between luminous objects and the technology that captured the light wave. Consequently, ultraviolet would be the closest light to us presenting with the shortest waves while objects presenting as red are said to be farther away. It's an interesting analysis when one considers that the first color of light on the solar spectrum that is visible to the human eye is red. At the highest end of the spectrum we find violet. This is a well known scientific fact. Bearing this in mind would we not expect objects that are red to be physically closer to us and those displaying violet light to be farther away? It seems logical. In the standard model of redshift motion of a luminous object is known as either *proper* or *peculiar.* Finally, the third form of measuring redshift is called Gravitational Red Shift. This process measures the lengthening of light waves originating from celestial body as it moves away from and out of its gravitational field.

In the standard scientific model, large galaxies emit matter and light. In the electric universe model it is commonly known that galaxies also emit electromagnetic plasma energy. Galaxies eject these energies outwards away from their central nucleus through massive galactic jets. Many of these jet streams extend from both sides of large flat active disc galaxies in opposite directions. Some galactic jets are single sided. In either case, the nucleus is the ejection point. In this standard model, the jet stream moving towards us on the side facing us will have a lower redshift. The jet stream on the opposite side that is further away from us will have a higher redshift. This commonly known model is accepted truth in modern science. Colossal galaxies emit massive streams of light and energy from their heart and divine center. In standard model science, the anomaly at the heart of a galaxy is a super massive black hole. In the electric universe we find a very powerfully charged energetic nucleus in place of a black hole. Ultimately, a galactic nucleus that is electric in nature resonates with the functions of the nucleus of the primordial cosmic egg. The standard model used almost exclusively to measure redshift of far away objects is known as Cosmological Redshift. This theory is embedded within the theory of an expanding universe. Wal Thornhill points out in the first of his prediction videos *Maverick Quasars and Redshift Values* that Cosmological Redshift only applies in a universe where expansion is absolutely unequivocally proven to be true. In his opinion, modern science has yet to provide definitive proof and factual evidence of this theory. Essentially, to entertain thoughts of this Nature did nothing beneficial for Wal Thornhills scientific career in the mainstream arena. Ultimately, in truth it is completely impossible for us to prove macrocosmic creation theories on any level. Nor can we observe the realm of the Absolute and come back to tell about it.

Where do Wal Thornhills alternative opinions against standard model concepts find their logical, intelligent origin? In his knowledge of the many scientific functions of the electric universe of course. When measuring the redshift of luminous, cosmic objects the standard models of science do not factor electricity into the equation. Furthermore, they do not take into account the numerous fascinating, observable phenomenon occurring all throughout our electric universe. In the electric universe models for which much proven evidence is endlessly being offered to modern science, the universe is not expanding. In metaphysics, the universe is not expanding. Our entire holographic universe is thought to be contained within the confines of the shell of the Cosmic Egg. In quantum physics as it relates to the seed of life, God is the dot or divine immortal center of the sacred sphere. Fundamentally, this is the original, primordial model of contained energy. These theories incorporate universal law into their construct and functions.

Sadly, there are multiple conflicting interests and tremendous obstacles to blending the two great minds of eastern philosophy and mainstream science together. If they could find some inexplicable, unexpected way to unite them, humanity would be subject to a blast of light so powerfully illuminating that it would blind us all. Massive explosions of light such as this are singularities that are very few and far between. That being said, the big bang theory itself may soon have legitimate competition. The standard model of an expanding universe is a bi-product of big bang cosmology. In fact, it is the foundational aspect upon which this origin theory is firmly rooted. Modern science bases big bang cosmology on three main factors. The expanding universe, the Cosmic Microwave Background and the abundance of light elements in space. If we are to reasonably consider the various, endless electrical processes occurring at all times in Nature all around us, within us and in the universe on macrocosmic levels, a whole new picture begins to develop. What we see is a radically, alternative concept of an origin theory that takes shape. In this new light, there is no expanding universe to consider at all. This detail however, does in effect, remove all credibility behind the fundamental, foundational aspects of the big bang model. Herein lies the main conflict and most major obstacle.

Astronomer Edwin Hubble after whom the Hubble telescope is named, invented the Hubble Law upon which big bang cosmology is based. Upon observation and examination of luminous objects in space and spectral analysis performed by Milton Humason the standards of redshift were established. Through careful observation of luminous objects, Hubble concluded that a decrease in a cosmic objects bright luminosity indicated a higher or increasing redshift. This conclusive result was determined using the Doppler Effect to measure redshift. As previously noted, the Doppler effect can be used to measure sound waves and light waves. Both of these forces of Nature are one and the same Source energy. Hubble's use of the Doppler effect arising from recessional velocity paved the way for the theory of the expanding universe. The modern scientific community of the time nailed down the concept. As such, it was etched in stone. The case was considered solved and this concept is still a construct of modern science to this day. This model of redshift was then adapted to fit the paradigm of a big bang origin by moving backwards in time. Utilizing this mode of mental analysis Hubble was able to produce a set of conditions that met the standard requirements which supported the model of an expanding universe.

Because no one was there to see it happen, any concept we come up with as to the origin of existence will always be theoretical at best. We can guess at macrocosmic aspects of creation that nobody witnessed but we must depend on science to provide solid proof for such claims. Therein lies the difficulty with determining any theory to be solid cold hard fact. We cannot see extremities similar to a big bang explosion happening today in our world. In regard to my research, which is to find the spiritual correspondences of Natural Law within the realm of science, this presents as an inconsistency. However, what we can witness in Nature are the same sacred geometric aspects of creation repeating themselves all throughout biological creation. In addition, we can also discover and measure the force of electricity found all throughout our world. In the case of origin theories that no one on earth participated in or documented, none of it can ultimately be proven by modern science. Accomplishing this goal lies within the realm of impossibility. The big bang theory ultimately came into effect as a result of Hubbles theory of an expanding universe. That is how this well known standard model came to be a staple in our academic world of science. All of this, may potentially have even been based on a possible miscalculation, a potential improper method by which to measure redshift. To err is human. Hubble made the most rational estimation that he could and backed it up by further theory and that was that. Scientists have been essentially attempting to prove these scientific creeds along with those of black holes, white holes, dark matter and so on ever since. In most cases, nothing credible or substantially solid enough to put a stamp on has actually been produced in reality.

Yet on the other side of the coin, evidence to support the fact that we live in an electric universe is so abundant that we currently require a pick up truck to hold it all. The plate is full. I will provide Wal Thornhills reasoning that there may be obvious irregularities with the current method of standard model redshift in the chapters ahead. As we move along we will be entering the stellar cosmic mind of Halton Arp to explore his professional thoughts on the subject. Professional astronomer Halton Arp bore the title Galileo of Palomar. He was highly respected by his academic peers and numerous members of the professional scientific community. Everyone knows who Hubble is. Humanity is aware of the existence of the Hubble telescope which bears the name of the man who coined the terms *expanding universe* and *big bang.* What is less commonly known is that later on in his career, Hubble began to doubt his own expanding universe theory. He began to consider whether the results of his experiments correlating luminosity to redshift were actually due to the Doppler effect or to some other unknown cause. Halton Arp expressed exactly the same sentiment in his intelligent, extensive scientific research.

The Galileo of Palomar has produced an extensive amount of academic material all of which is currently being studied by modern astronomers. If the expanding universe were to be proven incorrect in our modern, advanced technological age of reason it would change the face of science. All it would take is a revalidation of redshift by way of measuring the redshift of luminous cosmic objects using new methods. As a result, the foundational concepts of big bang cosmology could begin to crack. Academic science could suddenly be forced in a new direction. Consequently, a scientific void would be created where old outdated theories no longer hold up under the pressure of emerging evidence. All standardmodel origin paradigms would potentially fall into the category of outdated inefficient systems. It would become known fact that the current standard scientific models of an expanding universe and a big bang origin do not correlate in any way to known Natural Law. The academic void that would be created in effect, already has a host of fresh observable, provable concepts to fill it up. But this whole hypothetical concept, is a far reaching scenario to contemplate.

Through Halton Arps observations and perceptions we discover that we are highly likely to find galaxies directly in the same locality as quasars with redshift values that differ greatly. In his life's work, this is exactly what he discovered and extensively catalogued. Many of his findings are shown in his book *Catalogue of Discordant Red Shift Galaxies.* Needless to say, these findings did strongly oppose the known standard redshift model. Producing his stellar literary publication was a very brave act of attempted restructuring of science on his part. These things never go over well. Scientists who express resistance or objection toward the pre-established standard models of science are always met with great ridicule and rejection within the scientific community. As a result, each provable, credible example that Halton Arp provided was labelled a chance alignment and received no validation as proof. Ultimately this outcome could easily have been prophesied. This is a predictable response. Especially in consideration of the fact that he was producing findings that proposed a significant difference in redshift which normally indicated stellar objects at great distances apart in space. In effect, the characteristics of the quasars and galaxies he studied and mapped were essentially defying the scientific laws of standard model redshift. At the same time, in regard to my metaphysical research, his findings are of significant interest. Why, you might ask? My interest in his lifes work is rooted in the fact that his discoveries perfectly correspond to Universal Law as we know it to be a governing agent of all macrocosmic processes of creation.

Upon close examination of the images provided, Arp discovered what appeared to be evidence of high redshift quasars existing close to large low redshift galaxies. To complicate matters even more, when one looks at these images with a logical, rational, reasonable mindset, you don't have to be a rocket scientist to clearly see what he is attempting to show us. Standard model scientists continue to insist that what we are looking at in the examples he provided is a trick of the eye or an illusion. Yet one can still observe these images through the eyes of Halton Arp. When we do, a whole new story unfolds. In response to all of this, the standard model scientists provided Halton Arp with their explanation of Gravitational lensing. They then dismissed his findings concluding that what he was optically observing resulted from this known cosmic phenomenon. Gravitational lensing is essentially a trick of the eye. It is a false observation resulting from a process of refraction and bending of light around cosmic objects in space. This conclusion terminated any connection between quasars and active parent galaxies. According to the standard model redshift theory we should never be able to see a high redshift quasar in front of a large low redshift host galaxy. Why does this pose such a problem? Because in the accepted standard model a higher redshift object must be much further away from and behind these lower redshift galaxies.

In great contrast, when we look at redshift through the eyes of Halton Arp, an entirely different scene begins to play out on the scientific stage. In great admiration of this brilliant, stellar outside the box thinker, we will pay all due respects for his life's work by looking deeper into his vast pool of wisdom. In this study, we will not disrespect the standard models of modern science. My goal in this literary work is quite simply, to objectively and subjectively explore credible scientific information in a non-judgmental manner to find quintessential correlations to Universal Law. Acquiring knowledge is what our cognitive, rational consciousness has been designed for. Only modern science has the ultimate power to get us out of this mess and shed some radiant light on any confusing, inconclusive data. Maybe we missed something. New information emerges daily. It's ok to take another look and investigate further. The work of Halton Arp provides many pieces of evidence that appear intellectually credible. As his discoveries are famously known to oppose accepted standard models of science we will tread lightly on hallowed ground. Ultimately he has provided examples of quasars and galaxies interacting, as well as what appear to be higher redshift objects in the foreground of lower redshift bodies. In Halton Arps opinion enough evidence was provided to topple the standard model redshift paradigm. The modern scientific community has yet to agree. Such an admission could greatly impair other standard scientific models such as the expanding universe. If this model were tested and debated, so too would that of big bang cosmology itself. The whole world of science could plummet into chaos and total madness could ensue. But that's only the worst case scenario. The James Webb Space Telescope is now actively searching the cosmos for hard core evidence of the functions of our universe even in this very moment. Humanity is ready and poised to hear some new profound truths about the nature of our cosmic home and stellar existence.

In regard to massive host galaxies, small cluster galaxies and quasars, cosmologists, astrophysicists, astrologers and astronauts have never physically seen any of these cosmic anomalies up close. Thus, no one knows for absolute certain that high redshift conclusively indicates objects at great distances away from us. Our scope of vision on this subject will always be limited to and dependent upon whatever form of technology we currently have access to. The calculation of high redshifts equating to objects at great distances is based on the theory of an expanding universe and this too is fundamentally an impossible theory to inspect, observe and verify. As a race we can only guess at the functions of massive macrocosmic processes based on limited tools and observational equipment. So too, in regard to creation theories we can only make logical, rational guesses. Science can ultimately help us make very educated guesses that can be further tested, experimented upon, proven and studied. Today many modern scientists respect the body of evidence accumulated by Halton Arp. Numerous scientific professionals have an interest in seeing his theories either proven or disproven through further research, experiment and study. Some of our guesses as to how all things came to be here can be highly accurate based on sound ancient metaphysical teachings that resonate through every generation and culture of our race.

Humanity already knows of Universal Law, the Flower of Life, sacred geometry, quantum physics, fractals, the golden ratio, Fibonacci sequence, sympathetic frequencies, the golden rule and so on. By comprehending these aspects of Nature we come to understand the great manifestation that we are part of and in effect, the work of our Source. By travelling this metaphysical path backward in time we eventually reach the vector field. The sea of infinite creative potential and the possible realm of the Absolute. Herein we find the cosmic egg containing the holographic universe that we live in. In the electric universe teachings there is no big bang event to interfere or contrast with Natural Law. In quantum physics what we find in its place is a metaphysical super structure of magnificent architectural design. Making rational intelligent sense of Arps body of work and his many findings, is greatly facilitated through a comparison of his work to what we know about our electric universe and natural law in combination. Herein we find some relevant quintessential correspondences that the Galileo of Palomar was also searching for.

Once humanity as a whole discovers the true nature of our existence as it corresponds to universal law in the world of metaphysical science our race is primed for evolution. A spiritual revolution cannot fully gestate until certain scientific primordial aspects of creation are recognized and acknowledged. Only in a new light such as this can the spiritual-scientific evolution become revolutionary by any standard. A coming of a new age of light has long been prophesied, as we all know. These things too, shall come to pass. As long as we continue to seek the light of truth we are on the right path. But narrow is this path indeed. It's not so easy to find it in the world today, but it is still out there. Along with light comes wisdom. With wisdom comes freedom from the bonds of ignorance.

As noted earlier, before his passing physicist Was Thornhill made three videos predicting what he thought James Webb's first images of space would reveal. This chapter is comprised partly of a review of his first video entitled *Maverick Quasars and Redshift Values* which you can find at www.thunderbolts.info. His opinions will likely be scrutinized by many standard model scientists for years to come. As such his theories will inevitably be discarded by modern science. Ultimately, mainstream science does not support electric universe philosophies. This is obviously due to the many contrasting and contradicting aspects of the vast body of evidence presented which is in outright disagreement with known standard models of science. In other words, there is unlimited abounding evidence to prove that we live in an electric universe. Also, that electricity can be found all throughout our biological world and within our own physical constitution. Scientifically attempting to validate and prove this beyond the shadow of a doubt, clearly opens up the wrong can of worms. In fact it opens up that crazy can with all the coiled compressed snakes of science that come flying out at your head when you unsuspectingly pop the top off. The dreaded alternative science can. Wal Thornhill, Immanuel Velikovsky, Nicolas Tesla and Halton Arp are just a few members of the modern scientific community who took the risk of opening that can. Excellent analytical minds such as that of researcher David Talbott are also instrumental in helping us to connect the spiritual and scientific dots. These radiant under appreciated minds of our time continued to pursue their scientific goals despite the numerous consequences and setbacks they faced in their scientific careers. They were brave legends of the world of science whose courage and quest for truth in the face of adversary took more strength and resilience than most people even know exists. Fruits of the spirit such as this, which I admire greatly, are born of light, truth and endurance.

 Some may argue that their efforts got them nowhere in their search for Light. To believe this however would be to miss the point of their existence entirely. They were seekers of knowledge. Having found it, these wise ones fortunately left behind a massive repository of information, amazing science and alternative theories. Humanity still has the opportunity to crawl inside these great scientific minds anytime we wish to further investigate their incredible findings and discoveries. Ultimately these meticulously researched scientific studies deserve the same attention and recognition that all other credible, evidenced information is granted. Any logical, intelligent discovery is worthy information that is valuable, innocent and good until unequivocally proven wrong by proper scientific experimentation.

Wal Thornhills first prediction was that JWSTs findings

would support Halton Arps research. As the world of science is already well aware, Halton spent much of his scientific career attempting to prove to modern science that high redshift quasars are born in pairs and emitted in opposite directional plasma jets along the spin axis of a low redshift galaxy. In addition, Arp provided evidence indicating that as opposed to being located at the edge of the Universe, quasars optically appeared to be in the direct neighborhood of their parent galaxy. His academic dream was for the standard model Scientific community to embrace and further prove his findings and discoveries. To his regret, this vindication never came to pass. In harsh contrast, he was denied further telescope time at the Palomar Observatory for advocating against the accepted standard paradigms. Clearly, alternative concepts, outside the box thinking and radical theories have no place in the concrete world of academic science. In order to determine the accuracy of Halton Arps findings, further research is still required. If one touches on this subject what we find is that several brilliant professional scientists are busy doing two things. One scientific team will experiment upon Arps hypothesis, further research his work and then proceed to either prove or debunk his discoveries. This then gets followed up by yet another scientific team. If the previous results favor Arps findings, their conclusions are deemed inconclusive based on new research and data generated by the second team. So what we are seeing is an intellectual feud over what could possibly be deemed irrefutable evidence at some point in the future.

My little philosophy book is most certainly not the first or last literary work to speak of this famous subject and it will remain a subject of interest in the minds of many of the world's most brilliant scientists for years to come. Some will argue much like Wal Thornhill and many others, that results rendered by JWST present cases that conclusively support and vindicate Halton Arps lifes work. According to Wal, the first images from

James Webb do provide substantial proof of Arps claim that Cosmological Redshift may indeed be intrinsic. In essence, this would further serve to support and prove that

redshift is not due to an expanding universe. This concept lends credibility to electric universe teachings. In addition, Wal believed that the first Webb images appeared to lend logical, observable credibility to Arps theory of the origin of quasars by way of ejection from active galactic nuclei of large galactic bodies.

Now we arrive at the pinnacle of interest in regard to my research on this subject. As you know, in order to bridge the gap between science and spirituality, we must discover the quintessential correspondences between the Scientific processes of Nature and the forces of Universal Law. Both of these energetic constructs of creation are in operation in our universe, moving, vibrating and functioning all around us everywhere, within us and all throughout our planet at all times. Where are do we find a powerful nucleus conducting the many processes of creation? If you read the first book in this series, you already know that my path of study is of metaphysical context. As such, my conclusions encompass the known alchemical and eastern philosophical concepts of primordial creation. Wherein, heart of the Seed of Life is indeed a nucleus that emits energy or light. This model of creation also correlates to the operations occurring within the electric universe. The nucleus of the Cosmic Egg is the exact same force as the nucleus of the Seed of Life. These forces are one and the same metaphysical energy. This powerful nucleus, which in sacred geometry is located at the heart of the Isotropic Vector Matrix is the engine that generates the holographic universe. In these models, quintessential scientific correspondences literally abound. Studying the many procedures, scientific functions and processes of the holographic universe even further prove the presence of electromagnetism, plasma and electrical energy found all throughout the physical plane. Aligning these energies to the orderly processes of Natural Law, are the scientific dots that must be connected. The theory of the holographic universe and the Cosmic Egg are primordial metaphysical models of creation that fully abide by the orderly systems of Natural Law that govern our universe. Furthermore these origin of existence models can only efficiently operate in an electric galactic atmosphere. So what we discover is that Halton Arps findings and conclusions do correlate to Natural Law and the known metaphysical creation forces of our Source. Black holes on the other hand, do not. In the electric universe, the heart of a galaxy is a nucleus comprised of a pulsating plasmoid that emits electromagnetic plasma energy. Not a super massive black hole as indicated by standard model science. So bear this in mind as we proceed along in this study.

Halton Arp made the study of quasars the central focus of his life's work in astronomy. He photographed several cosmic samples of luminous bodies and galaxies. Through endless careful observation he discovered that many galaxies are often surrounded by quasars. He catalogued his many findings and produced his famous atlas of peculiar and disrupted galaxies entitled *Catalogue of Discordant Red Shift Galaxies.* Each of the samples he displayed had apparent adjacent quasars. He provided modern science with evidence which strongly indicated that quasars were found close to or near large central galaxies. He insisted that this was ultimately the case as opposed to being located millions of light years away near the edge of the universe. When you look at the image samples he provided this does indeed, optically appear to be the case. Despite this, his discoveries were rejected and sadly dismissed by the mainstream science community as always. This is a result of the obvious conflict of interest with current established standard models of science. In the year 2000 and 2009, Hubble released images of a group of galaxies called Stephans Quintet. The first set of images was taken with Hubble's wide field camera. Two images were taken with invisible light and the third in near infrared at 814 nanometers. In the following 2009 images, you see this breathtaking stellar galactic phenomenon along with a close up view of galaxy NGC 7319 located within this group. In 2022 Stephans Quintet was further explored and photographed by JWST resulting in even clearer and sharper images of these radiant galaxies.

Stephans Quintet Hubble 2009 NASA,
www.nasa.gov See images gallery, NASA

NGC7319

Hubble, constellation Pegasus, www.nasa.gov

Stephans
Quintet JWST 2022, NASA, www.nasa.gov see image gallery
NASA-TV

In the early 2000's Arp formed a research team with astrologers Margret and Geoffrey Burbidge. Together this well known science team discovered an ultra luminous object that turned out to be a quasar. The most amazing aspect of this discovery was the location of the quasar. It clearly, plainly appeared to be in the foreground of Galaxy NGC7319. It did not visually appear to be far behind it in any way as the standard model of redshift assumes. Halton Arp noted this discovery in his paper entitled *The Discovery of a High Redshift X-ray Emitting QSO Very Close to the Nucleus of NGC7319.* Realistically, Halton Arp and his team of colleagues were on to something big. This discovery was Amazing enough that it was ultimately recognized all throughout the modern, professional scientific arena. Today when cosmologists study images related to Galaxy NGC7319, Halton Arps paper listed above, is required research material. Obviously, if his work were total pseudoscience, this would certainly not be the case. Essentially Arps work is being used as academic training material despite its lack of credibility and acceptance within the standard model scientific community. According to Wal Thornhill, this fact lends tremendous, valid credibility to Arps research while causing standard model science to stand out as both supportive and publicly contradictory at the same time. Consequently, this is not a very open minded way to approach science.

By any measure it is clearly evident that his very significant discovery doesn't seem irrelevant to academic scientific advancement by any means. Galaxy NGC7319 has a redshift of $Z=0.0225$. It's local quasar has a redshift of $Z=0.2114$ which is much higher. The recorded redshift value of the quasar is 10 times higher than that of the host galaxy. This finding appears to prove the standard model of redshift based on distance to be invalid. Arp insisted that the Hubble image is showing us a very bright quasar in the foreground of the of galaxy

NGC7319. In addition, he observed a V shaped jet extending from the core of the galaxy towards the position of the quasar. This finding is of significant relevance. It may indeed indicate that there is an energy exchange occurring between the energetic nucleus of the active galaxy and its adjacent quasar. This demonstrates a high possibility of interaction between these two luminous bodies. According to the standard model of redshift this outcome is actually considered to be an impossibility. This finding has also been previously discovered and confirmed by mainstream astronomers taking radio wave observations. There is powerful evidence to support the near locality of the quasar to Galaxy NGC7319. This information is of vital relevance if we are to establish a legitimate connection between the two. Standard models of science place these luminous bodies vast distances apart in space. That being said, this evidence also serves to support Arps claims that quasars are ejected from the nucleus of the Galaxy as its point of origin.

At this juncture, it is also important to know how the modern scientific community explained the findings of Halton Arp, Margaret and Geoffrey Burbidge. As one might expect, the explanation supports the known standard pre-established models of science. Humanity was told that despite the obvious close appearance of the quasar, it was actually an optical illusion. This claim was supported by stating the standard code of redshift and insisted that the quasar was a background object. They further explained that it only appears in the foreground in the telescope images because of its great brightness. This brightness in effect was said to be outshining the cosmic dust and plasma of the giant disc galaxy. Bearing both schools of thought in mind one must examine the pictures for themselves to determine what the images evidently reveal. I have no interest in opposing the teachings of the modern scientific community in general but I will admit that the testimonial in regard to this specific example, requires fair reconsideration. There is still further evidence to bring to the table.

The 2009 Hubble images of galaxy NGC7319 within Stephans Quintet were taken with Hubble's upgraded wide field camera 3. These images showed the world an even clearer view of Galaxy NGC7319. Optically, the quasar clearly appears to be in the foreground of the host galaxy and not the other way around. It's position is indicated by the arrow in following the image.

Image NASA www.nasa.gov

Close up images of NGC7319 show a dark dust lane passing behind the Quasar and in front of the main active galaxy. If this dust lane is officially confirmed to be passing between these two anomalies it would provide adequate proof of a quasar in the foreground. Not millions of light years behind the galaxy as standard modern science has taught us. Evidence of this Nature would also provide a solid foundation to further support Halton Arps claim that the origin of the quasar is ultimately the active nucleus of the galaxy. Essentially, further examination, proper analysis of this image and a little flexibility on the subject could eliminate all questionability of the redshift controversy that Arp proposed. Is the quasar in the image really in deep space? Do your eyes tell you it's a trick of light or an illusion? Right now, the perceiver is you, what do you see? Is what we are seeing truly an optical illusion as the standard model suggests or is there more to this intriguing story? We have been told by the modern scientific community that the quasar may appear close, but this appearance of proximity is due solely to the bright luminosity of the quasar. Furthermore, the brightness of the quasar outshines all cosmic dust thus appearing closer to us by way of optical illusion. At this point, it's up to the observer to make sense of this argument. Many brilliant scientific minds are knocking heads in an attempt to intellectually resolve this issue. Various professional scientists and genius academics continue to view it as unresolved and inconclusive. Rightfully so. Humanity deserves a more rational, reasonable explanation than the issue being quickly boiled down to *smoke and mirrors.* It may be necessary for standard model scientists to simply take a second look at these images and conduct further research into this particular case. JWST and its amazing exploration teams have given us a birds eye view of Galaxy NGC7319. It's structure and functions are beyond spectacular. The research into this magnificent cosmic phenomenon is far from complete.

Galaxy NGC7319, showing dust lane and quasar. Jane Charlton, Penn State, HIST, NASA, ESA

In the standard model of redshift NGC7319 measures at Z=0.0225 which appoints it to a look back time of 311 million light years when the light captured by Hubble was emitted. The quasar with its redshift value of Z=0.2114 measures a look back time of 10.58 billion light years. As you can clearly see the standard model of redshift is placing the quasar at a distance of 34 times farther away than NGC7319. In regard to sorting out the confusion for ourselves, look at the images and logically decide what you see. From my perspective, I see what Halton Arp could see. Wal Thornhill also saw this galaxy and its adjacent quasar through the eyes of Halton Arp. The many contradictions associated with this paradoxal discovery could very easily be resolved by modern science and either put to rest or brought to fruition. There are many professional scientists who agree that as of yet there is no definitive intelligent or logical answer to this perplexing riddle aside from what appears to be obvious. It may be the case that we are actually looking at observable proof of Arps claims and perhaps not necessarily an optical illusion at all. The image above requires further research. It is magnified enough for simple independent analysis to be conducted. Examine the image for yourself and let your own eyes be the judge. Many scientists do indeed agree with Arps conclusions regardless of the lack of accordance between his evidence and standard model theories. All too often professionals in various fields of science hit dead ends with abstract new discoveries. All evidence which falls short of standard model theories will always get discarded or rejected.

The standard model scientific community lends no credibility into further research of outside the box concepts or radical new discoveries. In regard to exploring new theories, this approach is quite limiting and unfair. Academic brick walls, slammed doors and unmovable, stifling mental restrictions is truly not the foundation that real science was destined to be built upon. Would it not further intellectually evolve our race if we had the freedom to explore every credible discovery no matter how abstract? When heavily evidenced, should we not be exploring every potential scientific truth in existence? Sadly, this is not how the system works. Herein lies the tragic consequences and conditions holding astrophysics, cosmology, astronomy, quantum physics, plasma physics and humanity back, from a great global awakening. How can the sharp red eyes of James Webb contribute to an immanent spiritual evolution? By showing the world the truth. By delivering evidence and proof that will initiate a new scientific frontier of light. What academia lacks desperately, are modern scientific teachings and truths that resonate with universal law and the known scientific functions of our electric universe. Therein lies the bridge that serves to seal the gap between science and spirituality. Without conducting further cosmological research into fresh new abstract findings be they radical or not we will never spiritually evolve as a species. Our human race will only remain blinded by cosmic space dust. Vindication of Halton Arps work offers truth, light and hope. How so? His evidence shows us science that resonates with the law of correspondence. Correlations such as this, do effectively bridge the gap between science and spirituality. To find the truth, the scientific global exploration teams of James Webb, must look beyond the limitations of the current standard models of science. You never know when one might find a luminous stellar object hiding behind them.

In the world of science there exists a possibility that the official standard models of redshift, an expanding universe and the big bang theory could be flawed. Metaphysically speaking, these models do not correspond to natural laws of creation that we can witness occurring in Nature all around us. On another note, the electric universe teachings literally abound with validation of the primordial governing processes of universal law. This quintessential connection is key. Halton Arps work is rooted in knowledge that our universe is indeed filled with electrical processes occurring at all times all throughout the physical plane. The time for explanations rooted in reason is upon us. Under the right conditions the inquisitive, highly educated minds of todays modern scientific community can be raised to the status of heroes who usher in an era of logic and light. Science could be placed on a whole new pedestal. In an age of reason, all scientists have the potential to become superstars. Those who bring the ultimate light, such as Halton Arp and Wal Thornhill should be labelled as such. Legendary philosophers who think outside the box, deserve to be acknowledged and appreciated just as much as the next inquisitive scientific explorer. All great scientists who dedicate their lives to investigating the mysteries and functions of our universe should be rewarded for their wisdom and commitment to scientific light. Not excluding the Galileo of Palomar himself. Heralding in a scientific revolution will require some effort. Initially, there must clearly be a bending and flexing of standard model scientific law. A galactic exposé of truth can help initiate the long awaited global, spiritual awakening which is ultimately right on schedule.

In the next chapter, we will be looking further into Wal Thornhills JWST predictions at the second Thunderbolts video he created. In this video Wal continues his review of new information being sent back to earth with ongoing focus on the life work of Halton Arp. It is entitled *Cosmic Dust Proves Arp Invariably Right.* This video can be found at www.thunderbolts.info

CHAPTER 5

Attempting To Brush The Space Dust Off Local Quasars To Find The Visible Light

It is important to understand the role that dust plays when observing cosmic objects through a telescope. Often, dust in space can be opaque to visible light. This causes cosmic anomalies found within dust clouds to be nearly impossible to see and observe. Light in the visible spectrum can be blocked by clouds, dust or smoke here on earth. In the cosmos, dust acts in the same way scattering light between the luminous object, an observer here on earth and the technology receiving it. Infrared light is the best tool to use to penetrate opaque cosmic dust that blinds the light of luminous objects in space. Infrared light is capable of passing through various chemical compounds that comprise cosmic dust. This permits the observer here on earth to observe and study objects of interest unhindered by any visually obscuring matter. Modern science has discovered many different types of opaque dust clouds in space that block the light of background stars. Images taken of these cosmic anomalies in the visible light spectrum do not penetrate the dust enough to permit proper observable analysis of stars in the Cosmos. However, by using the infrared spectrum to view luminous objects in space our optical observational capacity is greatly increased. Dust clouds are penetrated by infrared for the most part which presents limited obstacles to investigating the once stellar objects and luminous bodies.

Scientists know that the universe is filled with multiple types of matter that can blind a telescopes vision. It affects what the telescope can see dependent on its viewing capabilities, its level of optical technology and the type of light it utilizes. In any case, the most efficient method by which to view and study our universe is through the infrared spectrum. James Webb Space Telescope, my most favorite eye in the sky, has been built by the worlds most brilliant engineers using the latest, most advanced technology on earth. To the great advantage of humanity, James Webb sees in red. These details make this priceless piece of magnificent equipment a valuable, vital asset to the evolution of our race. What we have now, is a very significant set observer scanning the vastness of space. Particles that comprise space dust are found all throughout the cosmos. These cosmic particles can have vast chemical compositions. They can consist of carbon molecules, silicone, water, hydrogen, ice and various other minerals and compounds. These various particles form the many dust clouds that we see in space. These great dust clouds present endless obstacles that inhibit telescopic observation. The particles of matter found all throughout the universe come in many different shapes and forms. They can be hard or soft and are normally very tiny and limited to a size of less than one micron or 1000th of a millimeter. The particles are microscopically small. To our human vision particles of this size are completely invisible.

Infrared light radiation has the ability to pass through dust particles because of its longer wavelengths. Infrared wavelengths are longer than those of the visible light spectrum. The wavelengths of the visible light spectrum are the same size as the many tiny dust particles found in space. In the far infrared spectrum cosmic dust can often produce a visible glow or a halo effect. The most famous reflection of this kind is the constellation Pleiades with its glowing blue halo around each of the nine stars that comprise this constellation. All 7 sisters and their 2 parents bear this strikingly beautiful stellar characteristic. In mid-infrared light, cosmic dust is visible. This is a result of thermal emissions, the heat it gives off. The heat emitted from the space dust is actually what produces and causes the glowing effect that we see around the stars of the Pleiades. These 9 stars are also of great spiritual significance in metaphysics. This open star cluster located to the north west of the constellation Taurus is also known as Messier 45. The famous group of stars is an immortal time marker on a great celestial clock. The ancient time keeping method involved with celestial cycles is similar to a Mayan long count calendar system. Both systems render accurate results over long periods of time and were of equal significance to our ancient ancestors.

When examining the quasar that appears to directly coincide with the dust Lane of galaxy NGC7319 shown in the previous chapter, the JWST mid-infrared MIRI images clearly show the evident dust lanes. In Webb's near infrared NIR CAM images the dust lanes are transparent thus invisible to the observer. This is how the amazing technology of the James Webb Space Telescope operates in contrast to other telescope technology created in the past. The JWST MIRI instrument captures the mid-infrared light emitted by the dust itself. On the other hand, Webbs NIR CAM captures near infrared light that passes through dust clouds. The image of Galaxy NGC7319 shown in the previous chapter, was taken by Hubble in 2009 using the visible light spectrum. Hubbles technology filters light at 814 nanometers in the near infrared. Hubble does not have the same infrared viewing capabilities as the more advanced JWST technology. The Hubble images did however successfully serve to clearly demonstrate that the quasar is comprised of *Bright Invisible Light*. James Webb is highly equipped for superior space exploration. Images of NGC7319 were taken by Webb using both cameras. In the images the dust lanes presented as a glow but were otherwise invisible.

The MIRI image showed the dust lane as glowing light all around the quasar. As Wal Thornhill points out in his own observations, every image shows the presence of dust and demonstrates how dust can block the visible light of a quasar. This is exactly what he indicates is occurring in the earlier Hubble images of galaxy NGC7319. Mainstream scientists have replied to this paradox by officially stating that the visible light of the quasar was coming from an extreme distance of billions of light years behind NGC7319. In the Hubble 2000 and 2009 visible light images, both the dust lane and the quasar can be clearly seen. To the naked eye they do indeed visually appear to be within the same general vacinity. However, this conclusion does not validate the official explanation of this anomaly offered to us by the mainstream scientific community. It is critically significant that their reply support known standard models of science. Therein lies a critical problem that has divided some of the most brilliant minds in the world of science today. Wal Thornhill examined these images and came to exactly the same conclusions. The modern scientific community made no further inquiries into the issue as they reply to claims such as this by reminding us of the standard accepted, firmly established scientific models. Which in this case, is that what we think we see, is only an optical illusion generated by the scientific process of gravitational lensing.

This is a hot subject in the academic professional world of science where we have two great schools of thought analyzing this data from all angles. While an official answer has been provided, the concept of an optical illusion was not accepted by all or determined through further study to unequivocally fall into the category of proven solid fact. As a result, the debate continues on the subject which is by many, still considered to be unresolved. Many scientists have a taken a major interest in this particular aspect of the scientific pool of knowledge. It certainly is a radiant subject of interest for many researchers like myself as well. Very soon you will understand why. Ultimately when we actually research the work of Halton Arp, what we discover is a system that corresponds to universal law and also to the many functions of the electric universe. These systems support one another in his analysis of quasar galaxy interaction in the most spectacular, reasonable and logical way.

The explanation provided by standard model science is that the quasar is ultimately in the deep space background outshining the whole disc galaxy including the dust lane. This very same dust lane can clearly be seen adjacent to the quasar and near the nucleus of NGC7319 in Hubble's visible light images taken in 2000 and 2009. These images are shown in the previous chapter and can also be found on the NASA website at www.nasa.gov . The Hubble images seem to clearly show what appears to be a very bright quasar in the foreground of the galaxy. The dust lanes do appear in plain sight to be positioned behind the quasar and in front of Galaxy NGC7319. With evidence such as this one might insist that such proof is difficult to deny and that such conclusions are easily drawn. At the same time the official explanation provided to the world simply states its analysis on the known standard redshift model of quasars being at great distances away.

Consequently, we know this answer doesn't suffice for the many professional astronomers and cosmologists who do concur with Halton Arps analysis. In fact, there are several professional scientists who agree with what he was so desperately trying to convince the academic scientific community in regard to galaxy formation, quasar-galaxy interaction and his theory of intrinsic redshift. As the world of science is already well aware, his theory of intrinsic redshift was outright rejected based on the standard model of cosmological redshift and the model of an expanding universe. Halton Arps claims contradicted, worked against and fundamentally clashed with both systems. Proving his work, would have ultimately debunked the known standard model of redshift. That is one really big reason why he was forbidden to pursue the study any further. What was the penalty? As the world well knows, Arp regretfully lost access to further exploration time with his favorite telescope at the Palomar Observatory. Why? Because instead of sticking to the program of the system, Halton Arp just couldn't stop exploring the abstract realms of science. He was fascinated by science in his expansive search for all intelligent possibilities out there. Open minded space exploration was his ultimate mission. In truth, that is what real science is fundamentally about.

Halton Arp could have lived for another 100 years and he likely still would not have been able to resolve his conflict of interest with the standard model scientific community. If *he* couldn't, its likely no one else can either. In regard to my own metaphysical and philosophical research into this subject matter, my aim and spiritual goal is to discover a set of logical, realistic, cosmic conditions that resonate with Natural Law as we already know it. This provides the only atmosphere in which to properly correlate the scientific processes of Nature to the Law of Correspondence. Some of the worlds most intelligent, highly educated, far reaching minds can be found within our modern scientific community. However, abstract thinkers in the professional scientific arena who oppose the known models of the system are up against a massive force. Cosmologists, Astronomers and astrophysicists are all obligated to follow a strict standard creed. From their perspective the quasar that is visually within the close proximity of NGC7319 has no other option but to remain classified as a luminous body located in deep space millions of light years away from its host quasar. Mainstream science will only continue to offer us academic evidence to support their position of the quasar to match their professional opinion and explanation. For scientists like Halton Arp, Wal Thornhill and Velikovsky, gaining scientific validation from the system, and the standard model scientists was an endless battle that they never won.

In the academic scientific community, all paths of knowledge will only lead back to established standard models of science and nothing more. Educators will continue to be told that despite what appears to be clearly obvious to the naked eye, the quasar is absolutely not in front of its host Galaxy NGC7319. In text books the quasar will forever remain a background object shining through the thick dusty discs of the large galaxy and all other space dust in between. By admitting otherwise, professional scientists demonstrate opposition towards the standard model of redshift, big bang cosmology and the expanding universe theory. Taking any opposing stance against known models of science would do nothing to advance their scientific careers in any field of modern science. They would potentially lose access to phenomenal research tools like Hubble and JWST and receive no funding for their proposed abstract studies. Thus, the proverbial index finger is held up to their lips in obedient scientific silence. Which is unfair. The law by which all mainstream professionals must abide is an unspoken code of conduct in the professional scientific arena. All explorers who dedicate their lives to studying science are aware of this limiting inconsistency in advance. In effect, falling into line with Halton Arps thinking would not be the best professional path for other scientists to follow. Even if their own personal observations and system of beliefs lends intelligent credibility to his findings.

Wal Thornhill recognized this great obstacle as a road block on the path to truth. In fact, he labelled the limited, frustrating condition created by this academic road block *a textbook case of paradigm paralysis.* Clearly, he felt very short changed in regard to acceptance of the same limitations that were the ultimate cause of rejections relating to his own lifes work regarding the electric universe model. Wal Thornhill, much like Halton Arp fought against the strict confines of the system for the duration of his entire scientific career. Yet, both men got nowhere in their efforts of gaining further funding for their heavily researched discoveries or validation of their highly credible, revolutionary findings. Despite this set back, these determined scientists stayed loyal to their academic goals, their research, evidence, theories, ideas, concepts and core program. Each of them shared a great hope of someday shedding light on essential truths that could illuminate the minds of humanity and the modern scientific community alike. In combination each of them bravely navigated a path of logic and reason which few men before their time have dared travel. Their tremendous efforts to have their work acknowledged and recognized within the standard model, professional scientific arena may have been to no avail. They did however gain the respect and silent support of many of these very same mainstream professional scientists. Yet, one would rarely discover this fact being expressed on any commonly known, public scientific platforms. On the same token, as a university student learning astronomy or cosmology you will find Arps work regarding redshift and quasar galaxy interaction has become recommended study. Ironic, isn't it?

Did Wal Thornhill or Halton Arps discoveries and entire body of lifes work ever amount to enough substantial proof or undeniable evidence to even budge standard model science from its lofty secure platform? No. Mainstream scientists do not embrace change of this sort on any level. However, this predictable rejection does not mean that their thorough, well documented research was incorrect, null, void or easily dismissed by other professional members of the modern scientific arena. It does however indicate that both of these scientists much like many who came before them were up against a restrictive force far greater than themselves. It was the force of authority, money and a predefined scientific creed with its own power structure. Regardless of this road block, they dedicated their lives to conducting their studies. Their gifts of immense wisdom and absolute stellar light have been left for humanity for further research. Many intellectuals on earth today from various fields of academic study view this gesture and their causes as great acts of kindness. In effect, their wisdom has now also become absorbed into the immortal collective mind.

It is vitally significant that any seeker of truth, light, spiritual evolution or transcendence of consciousness fully comprehend that developing your own system of beliefs is your own initiative. What you chose to believe in, accept or decline is always your own prerogative. It's important to investigate all relevant, credible information that you have access to on any given subject of interest. In cases where good data appears to contradict other good data and as such remains unresolved, research both sides and all facts carefully. Use your own rational, cognitive thought capabilities and logic to determine your personal stance on the subject. You own that freedom of thought as a human being who has been granted free will by your source. Once you've reached a mature, responsible age of reason, only you can establish what is ultimately right or wrong, acceptable or unacceptable, correct or incorrect, fair or unfair, ethically or intellectually just or unjust and true or false to you.

At this pivotal juncture it is imperative to note that many other scientists and professionals from various academic fields of study have made similar attempts to get new ideas across in a close minded system. They too meet with similar academic rejection, ridicule and intellectual sabotage. Radical thinking, outside of the box findings, unusual evidence and overlooked facts are rarely taken seriously by the academic world of standard accepted knowledge. This vice has a tight hold all through university levels and beyond. Despite their determination and persistence many great independent legends of science are known to often encounter tremendous opposition in their fields of study. Its not uncommon for open minded thinkers to have very little success getting their ideas across to standard model professionals. This same stifling paradigm actually applies to all other fields of academic study and research as well. In response to Haltons work, we are essentially being told that the high luminosity of a quasar will outshine the dust and the objects around it including the massive disc galaxy itself. This explanation does not eliminate the fact that many cases have been observed where bridges of matter actually appear to physically connect a host galaxy to its apparent local quasar.

If cases such as these are proved correct, the findings actually defy the laws of accepted, known science. What Arps examples seem display are legitimate, undeniable, observable connections between the quasar and the galaxy. Proving the existence of a high redshift super luminous object in the same space as its host galaxy is considered scientifically impossible. This assumption is based on the model of the expanding universe and big bang cosmology. As a result, Halton Arps concept of *Intrinsic Redshift* was quickly thrown out, instantly discarded and firmly rejected despite its high probability of legitimate accuracy. The concept as such was never conclusively proven or disproven by the standard model, professional scientific community. His theories and evidence were rejected as functional models with no further study conducted on their part at the time. Regardless of this massive, unsettling professional set back, Halton Arp remained convinced of his findings and continued to insist that quasars are indeed within the vicinity and local neighborhood of large disc galaxies. This information was of vital, significance as it provided a new platform from which to investigate the origin of quasars. Ultimately, Halton Arps main professional goal, was to show humanity and the world of modern science that the quasar is born by way of ejection from active galactic nuclei. In other words, the quasar is born by way of ejection from the heart of a massive galaxy. In addition Arps work also indicates that these quasars then proceed to evolve into galaxies through various stages of quantized redshift. In other words to break down or decrease and separate quasars by way of orderly natural fractal processes which ultimately alter the quasars original redshift. Understanding how all scientific processes of nature are governed by Universal Law, show us how all things in creation ultimately connect. In this book, I am seeking correlations that bridge the gap between science and spirituality. As such, to develop an understanding of the birth of quasars as suggested by Halton Arp is highly relative to this particular metaphysical course of study.

Both scientific schools of thought present their cases very powerfully. Many intellectuals from all walks of life and academic professionals alike, consider this mystery to be far from solved. However, in the minds of Wal Thornhill and Halton Arp, no such unresolved issue ever existed. From the perspective of the academic standard model scientists, the case is closed. Where do we go from here? We've reach an insurmountable academic obstacle. In order to make sense of all this confusion we must look a little further into this enigma to find out why Halton Arp and Wal Thornhill as well as many others are in total rational agreement on the subject. At the same time, it is also significant to comprehend that professional scientists employed within the power structure of standard model science are under strict obligation to agree with official standard model explanations. Always, and at all times. If they want to keep their jobs as scientists that is simply the price they must pay. Study material presented in in this chapter was an expanded review of the 2nd of Wal Thornhills three JWST prediction videos he created in 2022. For further research into the subject, his Thunderbolts video, *Cosmic Dust Proves Arp Invariably Right* can be found at www.thunderbolts.info .

CHAPTER 6

Examining The Big Bang Theory For Correlations To Natural Law In An Electric Universe.

In 2022, JWST took its first deep field image using its NIR CAM instrument which takes images in the near infrared spectrum. This stunning image was taken at different wavelengths of light over a 12.5 hour timeframe. NASA tells us that it took two weeks to produce the deep space image. The photos instantly became famous. Webb's advanced technology and infrared viewing capabilities far exceed those of the Hubble telescope. The deep space image focuses on a huge galaxy cluster. You can see a whole cornucopia of stellar objects scattered all throughout the image. Gravitational Lensing is a term used by the modern scientific community to describe the manner in which light bends around objects in space. This process distorts the observers view of the celestial body. In the deep field image luminous curved objects can be seen all throughout the image. These are said to be distant galaxies that seem brighter and larger which is an optical illusion. According to the mainstream scientific community, the visual distortion that we see is an effect of Gravitational Lensing.

This magnificent image shows us a snapshot of space the size of a grain of sand. This tiny space alone is filled with luminous objects and galaxies of various shapes and sizes. With all this activity going on out there in space it is logical to surmise that it is impossible for us to be the only human race in existence. Especially in a divinely created, very intelligently designed universe where all systems of creation infinitely repeat themselves all throughout Nature. In the deep field image some galaxies are small, some are large, some appear close to us while others are farther away. What we can deduce for certain is that the Cosmos is literally teaming with energy, stellar objects, luminous cosmic bodies and endless infinite mysteries.

The curved objects in the deep field image are obvious and apparent. Many of these curved luminous objects show a distinct *C shape* which appears as a semi-circle or a crescent moon. The electric universe teachings explain this optical effect differently than the standard model of Science. In the electric universe, the luminous spectral effect is caused by light waves moving through a neutrino filled, electromagnetic plasma field. Wal Thornhill referred to this medium as a *Neutrino Sea Ether.* In metaphysics, this plasma field is our local common vector field of infinite possibilities. In the electric universe model the optical effect of visibly lensed or curved objects seen in the JWST deep space field image is created by a process called refraction. Refraction refers to a bending of light through a medium. In this case, the medium is not a vacuum. Instead, it is a neutrino filled electric cosmos which is much denser around distant galaxy clusters. In this one image alone, there is so much to explore, catalogue, demystify and reveal to the human race. Isn't the James Webb Space telescope a wonderfully marvelous invention? Yes it is.

In the standard model of big bang cosmology we are taught that the universe began expanding after the Big Bang event. In the beginning, there was an initial brief period of expansion which is scientifically referred to as inflation. The process of inflation relates to the distance that the primordial jet stream of galactic matter and light that initiated the Big Bang would have travelled from its source of origin. In regard to inflation, the distance of the light given off by a luminous stellar object in space increases as it travels toward the observer. This effect is a result of the process of inflation and the expansion of the universe. As light travels toward the observer and the technology that captured it, space is said to continue to expand. In an expanding universe, once the light emitted from an object completes is journey to the source that captured it such as our Golden Boy, JWST, the luminous object that emitted the light will now be even further away from us in Space. *Look back time* is a term that refers to the distance that a light wave travels from the moment it is emitted from its point of origin while also taking into consideration, the distance arising from the expansion of space during its journey. This whole process is referred to as *co-moving radial distance.* Utilizing this system, all look back times are calculated and recorded. In this way space station observers can determine the distance of an object in light years away from us here on earth. The time that elapses since the light was emitted from a luminous original source such as a quasar is the actual look back time. It can also be referred to as light travel time. Both of these terms are used by cosmologists and astrophysicists to indicate the distance of high redshift luminous objects in space.

In known accepted standard model cosmology, we are taught that when a luminous cosmic object has a redshift of over Z=2 an optical effect is caused by the expansion of space. In contrast to the object appearing visually smaller the further it moves away from us as we naturally perceive it, the opposite effect will be caused. In short, we see it the other way around. In an expanding universe, when a luminous body such as a galaxy has a redshift value beyond Z=2 the observable size of the galaxy apparently appears to increase along with its increasing redshift and distance. So in other words the further it moves away from us the larger it will seem to appear. This is indeed paradoxical in comparison to how we naturally perceive objects at great distances away from us. We all know that when we optically view distant objects here on earth, the further away an object is, the smaller it will look. Herein lies the confusion. The whole issue may only be eliminated through expanding ones consciousness enough to acknowledge and accept the new and unusual. If the average person is flexible enough to give it a shot, perhaps someday NASA will bend with such flexibility as well. Just enough to embrace the sacred scientific nature of our existence a wee little bit. Out there in the cosmos, where the sky is the limit, anything is possible.

What does modern science need from NASA? A wider, more expanded mental search perimeter perhaps. A slight willingness to embrace credible non standard views would also be fair. What do I hope James Webb will find in order to validate my own spiritual research? I am seeking any direct correlation to a universal creation model that incorporates intelligent design, metaphysical science, natural law and the source of consciousness itself. Our Creator resides in the realm of the Absolute. In my research, that final plane may essentially be the vector field itself. In truth, the Big Bang is an ancient catastrophic singularity. A one time random event that accidentally resulted in the creation of our universe. This event cannot be observed repeating itself   all throughout Nature from massive macrocosmic to tiny microcosmic aspects of creation. On the other hand, intelligent design creation theories such as the ancient philosophy of the Flower of Life, though 1000s of years old, is still being taught in many cultures today. This model, as well a the metaphysical process of constructing Metatrons Cube and the Isotropic Vector Matrix absolutely do correspond to Universal law. In addition these geometric models of creation could not function in a universe that was not electric. So theories such as these serve to support the electric universe teachings which in turn, also reflect the many processes of Natural Law that are always at work in creation. The big bang on the other hand, was a one time event. Somehow, it occurred as a singular event in a universe run on systems and processes of creation that infinitely mimic, reflect and repeat themselves. Where does this paradoxal occurrence fit into a mystical world of science bound by Universal law? Think about it.

 Once you start seeing reality for what it is, it's hard to go back to being blind. Everything begins to click together like cosmic puzzle pieces. In a world built upon divine metaphysical order, every single thing in existence is connected through the spirit forces, sacred archetypal systems and scientific processes of Nature. Nothing in the world of Alchemy and metaphysics is considered accidental or coincidental. On a quintessential spiritual level, that rules out any random, irregular happening that resulted in the creation of all that we see. Our ultimate existence is due to the carefully directed consciousness of our Source generating various self maintaining, cosmic functions and spiritual systems. It all begins with a thought, five platonic solids, a ray of light and the seed of a flower. This information is key. It is imperative that future space exploration be approached from the mind-set of conscious intelligent design, and quintessential alchemical connectivity with a significant level of flexibility. These are critical key requirements of our scientific evolution on the eve of a great age of light. The divine omnipresent forces of truth and light are the heartbeat of the next generation. The human race is due for a spiritual, scientific revolution. This has been a long awaited event. We wait for it, because events of this Nature are encoded into our DNA. When it is time to wake up on a spiritual level, our inner self and higher self already know it. Just as naturally as it reminds us to wake up when we are physically sleeping.

Things that we optically perceive in our physical world, naturally appear smaller the further they are away from us as we all know. This quintessential pattern visually repeats itself all throughout creation. That being said, the concept of illusionary distance associated gravitational lensing does not necessarily correlate to any known common functions of Nature. Yet, it is a known standard creed of accepted science. Our universe is one massive clockwork system of scientific processes and patterns of creation that will always endlessly repeat themselves all throughout Nature. The random pattern of creation proposed by big bang cosmology and the resulting expanding universe model is known to be a one time origin event. This random occurrence and the conditions surrounding it, the causes and effects associated with this paradigm and so on do not repeat themselves in our observable biological world of matter. The scientific processes of nature that we can witness in action do not pattern themselves after big bang explosions.These current standard models of science do not account for divine intelligent consciousness. Nor do they resonate with the Law of Correspondence or the metaphysical processes of creation. Finally, big bang cosmology does not acknowledge or embrace the limitless electrical processes that are in operation at all times within our electric universe.

There is some serious intellectual opposition occurring between these two radically different perspectives. From a metaphysical vantage point the big bang origin of existence is spiritually void of omnipresent repetitive patterning. It offers no recognized reflective mirror of any known systems of microcosmic manifestation. There is no *As Above So Below* foundation to support this globally accepted standard model of creation. It demonstrates a lack of harmony with known immortal laws of Nature. Something needs to be done to rectify the spiritually deficient constitution of accepted standard model of creation. This is step one. In this modern time, a new age of light reason and logic are upon us. Humanity is more than ready to look to the stars to learn, acknowledge, comprehend and accept a new scientific truth.

Many standard models of science are potentially becoming outdated. Especially for those who are spiritually awake enough to see all the quintessential connections naturally encoded within all the things that we see in biological existence. From a metaphysical vantage point, the big bang does not incorporate into its design, the Isotropic Vector Matrix which is the engine and heart of the holographic universe. The architecture of the platonic solids and the critical role they play in biological manifestation is not even considered. These sacred archetypal energies embody the creative, intelligent forces of the Absolute and the divine consciousness found within all living things. We currently live in a technological age where Astrophysicists, plasma physicists, cosmologists and many other highly knowledgeable members of the modern scientific community are poised to become the next heroes of this generation. A lot is riding on JWST, NASA and their associate observational scientific teams around the world. Many independent, privately funded astronomers and cosmologists are also hot on the case. Who will bring humanity this long awaited light? All eyes can be fixed on the radiant stars and the splendor of creation in an instant, given good reason. The ancient well known spiritual significance of science must no longer be overlooked. All things small are a reflection of all things massive. All is connected above and below. All in creation is ultimately one. As are we ,as a species. This is the truth that we already know. This wisdom has long been confirmed by ancient magi, scribes, sages, ancient astronomers, ancient ancestors and other wise masters of light. This knowledge has been known for millennia.

Combined with valid irrefutable, verifiable, scientific evidence this wisdom becomes vital knowledge for the next generation. Truth is the heart beat of the new way in a new age of light. This consciousness shift is rapidly creeping up on us. As we usher in a global age of light, the objective is to keep it simple. It is beyond imperative that we de-mystify science. To initiate an awakening, humanity as a whole simply needs a reason *to care.* Our awakening race is truly in desperate need of less complicated scientific concepts that become are not so intricate that they cease to make sense to an average person. Once more it's up to James Webb to spot the illusions. It's up to the brilliant minds of todays scientific academic communities to decode the mysteries. The time to expose the truth about the reality in which we exist is now. The incoming age of light fixes all eyes on the cosmos and all focus on a new brand of spiritual science. Simply look at the current rate of growth and interest within the new age spiritual traditions. There is a movement currently on the rise. The spiritual revolution and the scientific revolution are *ONE* and *the same* energy. They are two forces that have always been united as one ultimate force of Nature at the heart of creation. This is the future. This is the way. This is the path and this is the light that we seek. This is the reward soon to come regardless of all other things that shall also come to pass. We are waiting for intellectual order to rise up from the chaos of the cosmos. The salvation we all truly seek is divine knowledge and spiritual wisdom.

The primordial systems of creation are naturally occurring consistent processes that have repeated themselves infinitely from the dawn of Genesis onward. We can find evidence of this divine omnipresent intervention all around us within every level of creation, above us, below us and within us. To truly comprehend the role that science plays in creation is to develop a relationship with the Absolute beyond ceremony and prayer through study. Even Christians can appreciate science when it is approached from the perspective of seeking to know how the creative work of God was conducted in the beginning. Anyone who understands the endless quintessential correlations connecting all systems within Nature can witness the creative forces of our Source in real time. The same macrocosmic processes of manifestation that occur at cosmic levels are always occurring in our physical, biological world and inside our material bodies every minute of our existence. Our biological and spiritual essence is contained within our earth vessel in perfect combination. Our bio-field energy is contained within oval and toroidal shaped geometric perimeters and electromagnetic concentric rings. Our solar system is even contained within a similar gravitational field. Furthermore, the holographic universe itself is contained within the confines of the shell of the Cosmic Egg. The consciousness of our Source is contained within the boundary of a divine sphere. Our consciousness in contained within the constitution of our mind, body, soul complex. In our biological world, nothing is endlessly expanding. Everything exists within its own energetic container. This vital detail is known to partially account for the purpose of the basket style container being held in the hand of winged Sumerian gods on ancient stone carvings. Even the energy that is found within a cell, is intelligently contained by an energetic membrane. All energy is known to be contained within a limited field, a barrier, an energetic boundary or a container of some form or another. As you can see, the endless limitless, nature of the standard model of an expanding universe does not correlate to universal laws of metaphysical, creation or our biological existence in any way. Even our sky which appears very limitless to our human scope of vision, has a limit. The vast seas may expansive, but even they are not limitless.

In the electric universe model, the universe is not expanding. Space is also not an empty vacuum. It is filled with electromagnetic waves of plasma and loaded with neutrinos. Additionally, the universe is thought to be of unknown age. When we study Halton Arps well respected body of work we find much observable proof and truth behind the evidence presented. Furthermore, we find several correspondences to Natural Law which is essentially what I am looking for. We can indirectly witness the process of galaxy formation and quasar interactions and understand it in comparison to human reproduction. The quasar is born of the plasmoid womb of the parent galaxy and ejected outwards into the world of matter, just like a newborn infant. Arps work shows us examples of where newborn quasars appear to interact with their host galaxy. New newborn quasi stellar luminous objects begin their life as high redshift quasars within the local area of their parent galaxy. Eventually they break apart forming smaller cluster galaxies through a process of evolution and fragmentation. These quasars eventually evolve into larger independent galaxies as they reach a certain distance in space away from their host galaxy. Several examples to support his theory extensively showing how this process occurs can be found in Halton Arps well documented book *Atlas of Strange and Peculiar Galaxies*. Halton Arps concept of galaxy formation and quasar interaction are reflective of and similar to human reproduction, childhood and teenage evolution to a point of independence in our human world. If human childbirth correlated to big bang cosmology in the same way, every delivery room would be a total bloodbath. The infant child would be exploded onto the physical plane in fragments of organic molecules. This mass of newborn tissue and cells would fly all about the room, eventually forming a human through a process of coagulation. The infant child would then continue to expand indefinitely. The planet would not be big enough to contain it. Clearly, this is not a healthy state of existence nor a safe way to enter the world. The concept of exploding infants that expand forever deifies the known laws of biological existence. Thank

God childbirth aligns to the intelligent Natural Laws of sacred geometry and quantum physics instead.

According to his knowledge of the electric universe, Wal Thornhill, expected the JWST deep field image to show us galaxies that get fainter and smaller. Some naturally bluer, or redder depending on the telescopes viewing capabilities, technical power or optical limitations. Based on his analysis of the image, this is exactly what we are being shown in this first deep field image taken by Webb. We see families of galaxies and clusters in just a small part of a massive universe of unknown size. There is clearly so much more to explore and discover that it is practically overwhelming to fathom. Science can take it one step at a time. With the successful launching of James Webb in 2022, we have only just began this wondrous scientific expedition of breathtaking findings and spectacular new discoveries. There is so much more to come.

Wal Thornhill firmly believed *that nature must be explained by physics.* He insists that mathematics only describes the result in a reality where physics isn't math and endless equations. In the electric universe model there are no collapsed matter objects. There are no white holes. There are no black holes. There are no neutron stars. All scientific phenomenon can be much more easily understood when one considers the many systems and processes occurring in the electric universe at all times. With a comprehension of the scientific processes of Nature the existence of the electric universe is made even more evident, apparent and intellectually digestible. This imperative detail cannot be stressed enough. Both Wal Thornhill and Halton Arp insisted that *Lensed Galaxies* that do not appear as a result of refraction of light through a denser neutrino sea, are in fact, newborn high redshift quasars. They further insisted that these quasars were, in actuality infants of nearby galaxies. If they were at such great distances apart as indicated by standard model redshift, scientists would not be able to see them so clearly, let alone examine their structure or characteristics. In contrast to standard model teachings the quasars are said to be nearby or local. Images provided by Halton Arp show one piece of evidence after another that visually indicates that this does indeed appear to be the case. He does bring plenty of credible evidence to the scientific table to support his claims. Observation and further examination of his data does support his theories, which ultimately appear to make logical sense.

At this critical juncture I feel it necessary to make note of the fact that NASA has recently put forth a model in an attempt to potentially explain obvious quasar galaxy interactions. Their unusual explanation involves cannibalism. The process offered describes a system in which quasars are *consumed* by parent galaxies. Um, ok. So here's what we are going to do. We are just going to leave that theory right where it is and stick to what we know about quantum physics and Natural Law. In order to embrace a theory of this sort in this particular study, we would be obligated to explore cannibalism as a natural repetitive process found within creation all throughout Nature. We don't necessarily see that happening in our material world all around us. It's actually a pretty gory, horrendous, destructive thought to consider. I will say that the Aztecs had a poor diet. Their cannibalistic traditions didn't make it too far into modern society. Behavior such as this certainly does not constitute any kind of process found anywhere within orderly communities. It is simply not natural. Cannibalism does not correlate to any sacred fractal processes of creation from a metaphysical perspective. As we move forward we will continue with a more humane exploration of the scientific processes of galaxy formation from the perspective of Halton Arp. Not to insult the modern scientists who devised this cannibalistic quasar theory in any way. My aim is to simply adhere to the goal of this literary work. To find quintessential scientific correspondences to Natural Law it is important to maintain the current direction of the research which I am conducting. Within Halton Arps well documented scientific life work we do find endless natural correspondences to the many known, intelligent orderly systems of creation that are governed by Universal Law.

James Webb's 1st Deep Space Field Image 2022. www.nasa.gov NASA

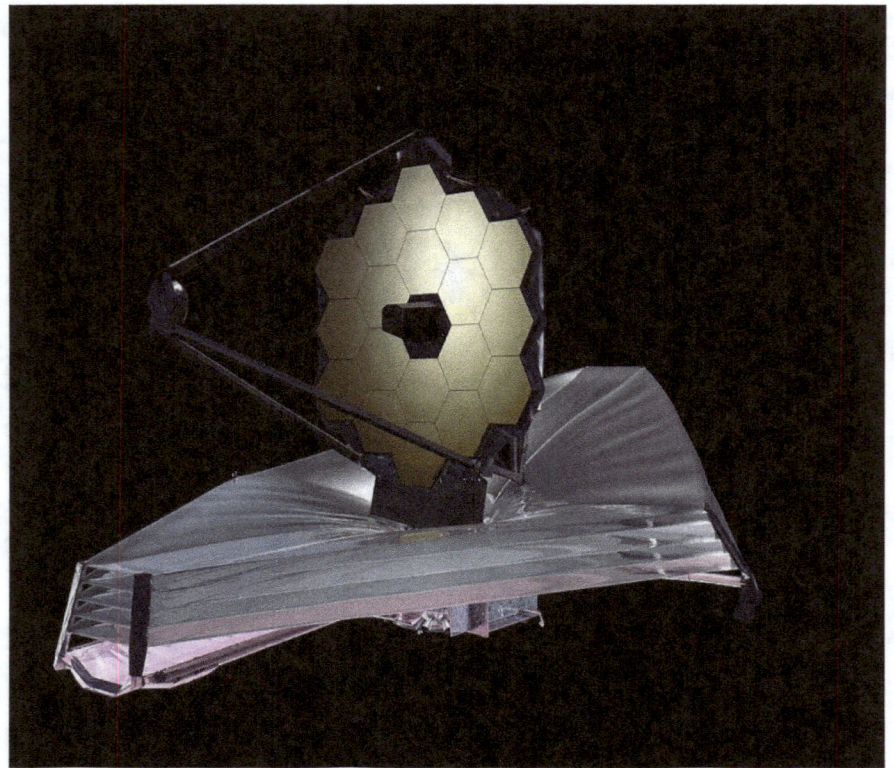

Launched In 2022, James Webb Space Telescope is a technological work of art. Image NASA, www.nasa.gov

CHAPTER 7

A PIVOTAL CROSSROAD. ARE WE ON THE BRINK OF A SCIENTIFIC REVOLUTION?

The time has come to embrace the fact that science is due for an inevitable evolution. With the advent of the 2022 launching of the James Webb Space Telescope, the world of science has exploded with new evidence and mystifying data to decode. Space exploration has stepped up the game. JWST combined with Hubble Space Telescope, are a force to be reckoned with. This amazing technology is causing the potential effect of the emergence of a whole new scientific frontier. Accepted standard models of science are being powerfully challenged. More so now than ever before. Even the big bang model itself is under the microscope of questionability. As further sound, intelligent information continues to emerge the potential now exists for new evidence to prompt new research into areas we once thought science had completely already solved. Initiating new levels of cosmic data collection brings new ideas to the table along with new solutions to old scientific perplexities that many members of the scientific community still deem inconclusive.

Many brilliant scientific experts from various fields of study are gently speaking out against certain systems that mainstream science has always taught to us as hard science based on accepted fact. As JWST continues to send back phenomenal images to earth, an abundance of evidence to support the orderly systems occurring within an electric universe is stacking up to the height of a mountain. This change that we all feel in the scientific air around us is of great significance in todays modern age. New technology brings forth new data and new discoveries are made. Sometimes these new discoveries usurp old models that have always presented with unresolvable inconsistencies. So be it. In many ways science is ready for a new perspective. The electric universe and its many orderly, observable systems and processes is rapidly getting exposed and gaining credibility. In an electric universe the known, accepted, standard models of science simply do not apply. Interestingly enough, we could be only a few discoveries away from a place where the science presented in the electric universe model can gain some level of precedence over old systems of belief. Or at the very least, earn some solid ground to stand on in the modern scientific arena. Even this would be a refreshing change in a world that currently ignores these intelligent concepts.

Steven Hawking, the famous beloved physicist was a huge supporter and promoter of the standard models of science as we all know. He was limited physically but this never stopped him from becoming a famous household name and a star in the modern scientific arena. In no way, shape or form was this famous genius scientist limited intellectually. Steven Hawkins wrote a best selling book entitled "A Brief History of Time". In it he explains the characteristics of a scientific theory. In his legendary opinion all physical theories are provisional. In other words, mutable or changeable due to the fact that they are ultimately only a hypothesis. Essentially, you can never prove a hypothesis. Even when the results repeatedly agree with a theory it is still unknown as to whether or not the next time the experiment is conducted, the results may be different and contradict the theory. At the same time, a theory can be disproven and disqualified by finding even one piece of evidence or one observation that disagrees with what the theory suggests and predicts. Philosopher Carl Popper emphasized that a good theory makes a number of predictions that could ultimately in principle, be easily disproved or falsified through experiment and observation. Each time the experiments are in agreement with the prediction, the theory survives. However like Steven Hawkins he also makes it clear that when any piece of evidence, experiment or legitimate result is found to disagree with any theory, the model is abandoned, discarded or modified. As you can see, theories undergo intense evaluation and scrutiny in the world of science. Sometimes however, the system can literally work in favor of a theory. Wal Thornhill believed that the evidence being brought forth by James Webb served to support the electric universe model. In the mind of Wal Thornhill, new information resulting from these findings lent great credibility to the vast array of electrical processes occurring everywhere in the Universe. In turn, the findings also appeared to lend less credibility to black holes, the big bang origin theory, dark matter and the existence of an expanding universe.

 Richard Feynman, the most intellectual physicist of our modern era also gave us a brilliant explanation of the scientific method. In a lecture at Cornell University in 1964 he stated that we first look for a new law in science by guessing it. Then we compute and deduce the consequences of the guess to see if the law that we guessed was right. Next we see what it would imply then compare our calculations to the scientific processes of nature or universal law, to experiment and to experience. We essentially compare it directly to observation to see if it is wrong. According to Richard Feynman, this process that is followed in the scientific method was the key to science. He stated that if a guess, theory or hypothesis disagrees with experiment "its wrong. That's all there is to it." Of course, he was exactly correct. As a result, scientific research and study are left open to a non-selective objective study of theories, possibilities and hypothesized models. All of which can be investigated in a non-bias atmosphere. If we were to follow the standard set out by Feynman in his Cornell lecture, it would change the foundation of what standard model mainstream science teaches us in regard to the big bang. This theory along with cosmological redshift as the reflection of an expanding universe would have to be re-explored, re-evaluated, potentially redefined and possibly rewritten. In practice however, standard model science does not work that way. Once modern paradigms are established these theories simply become fact. The proposed scientific models become mountains that are difficult to crumble.

Thomas Kuhn was the most influential philosopher of the 20th century. In his book, *The Structure of Scientific Revolutions* Kuhn explained the common restrictions of standard model science. In his opinion, with firmly established paradigms such as the expanding universe or the big bang theory the function of science becomes the exploration of the paradigm or model with which they have been presented. The scientists are to learn the model and continue to establish its credibility by whatever scientific means necessary. All research, experiment and observation naturally becomes centered on that goal. Ultimately this preset standard of established models, systems, paradigms and concepts is not to be questioned by those who pursue a career in science. Essentially this type of science doesn't leave room for questioning any area of scientific study. Mainstream scientists are set to the task of problem solving within the known pre-existing paradigm which they have been given to research. This method limits outside influences such as other points of view, alternate theories, radical concepts and fresh objective perspectives. Anomalies, abstract thoughts, outside the box thinking and so on are ignored, rejected or dismissed. Eventually, restrictions and limitations of this Nature can have a tendency to result in the gradual emergence of scientific revolutions. We are beginning to see the onset of this paradigm shift right now from within the professional scientific community.

Intelligent professionals from all areas and fields of scientific study are analyzing, researching and decoding the mysteries and anomalies of the cosmos as seen through the eyes of JWST. According to Wal Thornhill undeniable evidence is finally being presented that could potentially contradict the standard models of known science. Thomas Kuhn further stated in his book that the continuation of a scientific paradigm in the face of obvious contradictory evidence equates to a condition he termed *Paradigm Paralysis.* Clearly the condition that results from ignoring sound scientific evidence contributes to mental stagnation, obstacles on the path of reason and a limited intellectual evolution for our race. Thomas Kuhn wasn't wrong. A paradigm shift is imminent within the scientific community. An expanded focus of the Spiritual aspects of macrocosmic creation is becoming a critical necessity. A new age of light and reason looms on the horizon. Ignoring sound evidence from any valid scientific source can indeed be detrimental to the evolution of science as a whole. A non biased approach must be taken in order to reestablish a new universal cosmic reality that incorporates metaphysics and Universal Law. This force is already flowing through the heart of the new age spiritual world where science and spirituality merge within one community. If modern science was set to the task of decoding the secrets of sacred geometry in order to potentially expose the energetic matrix of our holographic universe, humanity would be on a fast track to expanded conciousness. We could potentially amalgamate all esoteric knowledge into one tested, proven scientific truth. What an illuminating notion. This energy is currently highly required within our biological world.Light such as this is the fabric of the new Aquarian age of reason, logic and spiritual ascension. All of it, is now. It is on time and long overdue all at once. Open your heart, absorb the light, reject the coarse and unsavory. Our willing participation in spiritual transformation is a key to success.

The accepted standard models of science as we know them are deemed academically indestructible and un touchable. Despite this, many intelligent open minded scientists have speculated upon alternate scenarios and alternative research that contradicts mainstream science. A limited scope of vision serves only to set scientific discoveries back as opposed to evolving all newly discovered theories and observations to the next level. Herein lies the core of the stumbling block that limited the vindication of the highly accredited life work of Halton Arp, Immanuel Velikovsky, Wal Thornhill, David Talbott, Nicolas Tesla and many other genius minds. Despite its heavily evidenced and well documented existence, the electric universe hasn't been offered any solid ground to stand on in the modern world of accepted science. This is because everything we know about our electric universe threatens the existence of the standard models of mainstream science. Do we know that atoms are electric? Do we know that water is a conductor of light and electromagnetic energy? Is our watery biological body electrical and di-polar in nature? Has our sun been exposed as an electric celestial body? Is there any electrical activity seen in stars and galaxy formation? The answer is Yes to all these questions. It is not that hard to find these electric charges all throughout our biological world of creation. Experimenting on electrically charged aspects of Nature like our own cells shows us the electrical functions of atoms and molecules under a microscope. Forces such as this surround us awaiting exposure by modern science as totally true and non taboo.

The known standard models of creation have stood their ground by casting out alternate theories since their inception. Yet, statements made by some brilliant scientific minds indicate an overall general lack of satisfaction with the mentally restrictive limitations surrounding the way science works in mainstream academia. Interest in further exploration of radical concepts new theories, alternative hypothesis or new models presented by scientists don't normally find their way into modern textbooks, established science manuals and university curriculum. Does this restriction mean that alternate theories never get researched in the world of science? No. Not in any way. In fact, independent scientists are always busy at work conducting various forms of study and research that is entirely independent of the standard model world of science that we all know of. Having established an understanding of the strict boundaries associated with modern science, you can easily see how abstract thinkers have such great difficulty breaking the shackles of the system. Regardless of how brilliant and credible their outside the box evidence may be. Surprisingly, even the inventor of the Hubble telescope himself, Edwin Hubble expressed some skepticism towards the expanding universe paradigm near the end of his scientific career. Today, with the new amazing optical abilities of James Webb on the job, more clues to help us solve the mysteries of our universe endlessly arrive in great abundance. In the eyes of several professional scientists and academic researchers this new information could even provide further proof that potentially contradicts standard model redshift. In order to validate the evidence presented by Halton Arp a reevaluation of redshift would be a critically mandatory requirement.

A university education in mathematics isn't necessary for those who simply wish to acknowledge and comprehend the basic physics of Nature. According to Wal Thornhill, mathematics is a descriptive tool to be applied to real physical models. He was known to state that physics isn't math and that mathematics only describes the results of physics. In fact he was correct. Even ancient alchemical and esoteric traditions taught geometry as the core of physics, not math. Basic fractal equations and platonic solids do not require three chalk boards full of equations to comprehend. When geometry and physics is complicated to such an extent, hardly anyone in the common population can make any sense of it. Essentially science is for every one. As such, physics should not be made so unnecessarily complicated that average people cant learn it or spiritually benefit from its righteous, essential teachings. In truth, ancient wisdom of the past could very easily be at the helm of a new generation of science exploration in a new Aquarian age of reason. From our instinctive ability to learn from the past we know that history always repeats itself. Bearing this in mind, humanity must come to acknowledge that the key to understanding the present and the future, always lies in the past. Understanding events and natural disasters occurring in the past can even help us to understand the cyclical nature of climate change and the atmospheric conditions existing on our earth today.

Wal Thornhill dedicated his life work to a study of the science presented by Immanuel Velikovsky. Velikovsky offered us a spectacular cosmos where brown dwarf stars like the planet Saturn eject cometary matter that evolves into planets, like Venus. His best selling book *Worlds in Collison* published in the 1950s is still being purchased today. In its day, this book made a tsunami of a splash in the academic scientific community. His abstract findings did not fall within the safe parameters of mainstream science. As a result, much like Nicolas Tesla the work was shunned and his evidence was rejected. Furthermore, Arps theories were never tested, experimented upon, proven or disproved the members of the scientific community that rejected them.

It's a well known fact that many researchers, archeologists, paleontologists, geologists, physicists chemists and so on eventually encounter boundaries of similar nature. Sadly this issue is incredibly common. Once more I suggest that my readers look deep into the past. This is where we find the answers to scientific questions still being asked today, regarding the metaphysical functions and operations of creation. This light resides there, as all things in nature continuously repeat themselves including historical events and cyclical processes of creation. Wal Thornhill honestly believed that if modern astronomers were to simply acknowledge Velikovsky's classic well researched and heavily documented book *Worlds in Collisions* it would change the face of science as we know it. He wasn't wrong about that. But did it ever happen? No, open minded approaches such as this would take a miracle. Velikovsky meticulously studied ancient creation stories of human antiquity in areas all around the world. David Talbott is another excellent modern researcher who spent a huge part of his academic career studying and further expanding the great scientific work of Velikovsky.

Talbott decoded secrets found within the writings of this amazing scientist and ultimately uncovered the same patterns of primordial processes and systems of creation being expressed within all global cultures. He intelligently applied our modern understanding of astronomy, cosmology, cosmogony, physics, logic, common sense and reason to these legends and stories. In this exact way, he solved an ancient puzzle that modern science has not studied, tested, experimented upon or validated in any way. Sadly, this is not an unexpected result. In light of this drawback, his theories were never proven or disproved by accepted academic science and as such, were never taught in schools. His whole body of work is still open to this challenge and still up for a good debate. David will always hold firm to his excellent belief system with all due reason. His work is filled with profound, enlightening concepts, answers to ancient rock art riddles, profound revelations and wisdom. It is of great significant spiritual relevance, that researcher David Talbotts work be reconsidered by modern science and explored by all rational minds on earth from every walk of life. His research encompasses an understanding of the profound spiritual aspects of Velikovsky's findings. Through his study of Velikovsky's work David produced a series of Thunderbolts videos which can be found at www.thunderbolts.info. His research is illuminating and very unique. His information is good. At the very least, let us all respect good work when it's done well. Thank you David Talbott. In order to solve the mysteries of today we must decode the secrets of the past. That is exactly what these human beacons of light mentioned herein, have attempted to do. Their scientific findings have been donated to the human race as great selfless, academic contributions. Their work has been dedicated to our intellectual, spiritual and scientific evolution. This is exactly how great teachers, researchers and brave explorers become historical figures in linear time.

Woven into the legends of the past are the gods of mythology. Zeus with his massive thunderbolt and Perseus riding the chariot of solar fire with its flaming wheels. Egypt has so called dark, evil beings such as Anubis and Seth. Christian religions have been filled myriads of demonic entities like Lilith, Satan and many other evil devil monsters that are believed to legitimately exist. In truth, such defeating energies have no place in Newtons clockwork planetary system. There are no devils in physics. In metaphysical practices the ultimate goal is to connect to higher self, to Source and to the oneness of the collective mind. The essence of our creator is unconditional love. Not evil. Yes the Law of Duality exists as does the Law of Rhythm. While both of these forces are in operation at all times in nature it is our own state of consciousness that dominates whether or not evil or hateful thought forms enter our sphere of existence or personal environment.

When you reach any true lofty level of awareness you come to realize that the devil is a construct of mankind. Not of God. The devil and demons are entities that were created by the religions of humanity to instill obedience in the minds of man. Demonic concepts emerged from religious, spiritual and tribal traditions through humanity's inability to explain certain aspects of Nature that they perceived as terrifying or bad. Volcanoes, lightning, floods, hurricanes, earth quakes, meteor strikes and so on were deified then appointed divine personalities and characteristics. Energetic forces of Nature are very real indeed but it is critical to comprehend that they only exist as personified beings within the thoughts of man. As such, humans can manifest these thought forms into energy which can ultimately come to exist at lower vibratory frequencies. This energy can be felt, sensed and sometimes even experienced by others who choose to exist in the same lower vibratory state of consciousness. In truth, the being called Satan that everyone is so terrified of only came into existence around 2000 years ago around the beginning of the A.D Anno Domini new Christian era.

Seeing as how its now 2023A.D, you can see that the devil is a rather new figure in world history. Thanks to the thoroughly complied, heavily researched findings of Freddy Silva, humanity can now trace our history and spiritual traditions all the way back to 15,000 B.C.E. Reaching this far back into ancient antiquity does not expose any devils. More interestingly, we find symbols of solar deities. Petroglyphs from this ancient time reveal a whole variety of sun symbolism and depictions of the functions of electrical energy. We see endless examples of the symbolism of the Flower of Life carved on ancient temple ruins found all around the world. This symbolism existed in human culture long before the rise of the great Harrapan Civilization of ancient India. Even before Petra was constructed. Additionally, in the symbols of antiquity we discover a recorded history of floods and global cataclysms. Often, the immense terrifying energy of these cataclysms would be personified into hypothetical deities. That is where the concept of these darker beings initially originated. Eventually, global religious traditions evolved these previously known personifications of dark forces into literal evil demonic devil guys that are still thought to exist today.

In reality, when you study the science of the electric universe, you encounter no devils. What so called evil we do discover, is the knowledge of global catastrophes, world floods, exploding volcanoes, deadly plagues and apocalyptic events that have occurred in the past. These dark forces do exist on the physical plane at lower levels of vibration. There is no information in the modern academic world of science that has even been able to explain catastrophes associated with the angry mythological gods of antiquity. We have been offered no scientific explanation for Noah's flood or for the 12 plaques of Egypt aside from that given by Immanuel Velikovsky in his best selling book *Worlds in Collision*. Yet, somehow these mythological stories stand the test of time and are recounted, recalled and retold within every human generation. Obviously, mainstream science has left out some very critical details about our existence. These stones, having been left unturned for so long were covered in the dust of ignorance, hidden, buried and lost in time. Then came the meticulously researched findings of David Talbot. Mythology as we know it, is all that remains of a very turbulent time in history a long, long time ago. Mythology is an echo of the ancients. Few of us are actually aware of the fact that these myths stem from actual interplanetary events in our ancient, distant past. They have less to do with respectfully fabricated multi pantheons of gods and much more to do with reality, ancient tribes and world history. Each God form that humanity creates is simply an energy that exists in Nature that has been personified so as to create a spirit form with which to communicate. It is a character created to represent a specific aspect of metaphysical energy found in Nature.

Catastrophic events equating to our concept of Armageddon have indeed occurred in the past. The exodus of Moses and the Israelites in the Bible was a historical account of such an event. Exactly what was happening in our local area of the Cosmos when catastrophic events and other divine acts of the mythological Gods recorded in the past were being enacted? This is the most relevant question to ask. The damaging activity in the Cosmos was actually visible and observable within earths atmosphere. Humans saw it. They recorded this on cave walls in the form of rock art. Images, petroglyphs and monuments were designed and erected so that no one would forget. Yet, lo these many ages have passed and do you see? We have foolishly forgotten these things as a race regardless. The events we witnessed in the sky, produced a crisis here on earth. The events were of such great significance that they were recorded by every generation who witnessed them. These global historical accounts were the exact same stories that became the myths and legends of the ancient gods that every generation of humanity knows of. This is what Velikovsky discovered and what David Talbott spent a huge part of his life expanding on.

This chaotic, cosmic scenario is exactly what is explained in Velikovsky's book *Worlds in Collison* from a theoretical, scientific perspective. This profound literary work is not for the faint of heart. It's for those who seek the alternative truth and wish to get to the heart of the matter. The information provided in his book is very valuable for the human race despite its lack of recognition by mainstream science. So skeptics everywhere, read it with an open mind I implore thee. *Worlds in Collison* was heavily researched. All of the information he provided still exists, is still testable and present to be experimented upon to either prove or disprove. I anticipate the dawn of a new day when the modern scientific community flexes and embraces the unusual just enough to enough to allow a serious study of Immanuel Velikovsky's entire body of scientific and literary works. However, the great obstacle of academic science being structured upon standard untouchable theories, will always prevent studies of this nature from being conducted in the modern scientific arena. Standard model scientists offer a whole host of reasons why this is the case which are ultimately all very valid in their eyes. Realistically, the chance of this permission being granted or the endeavor provided with funding are slim to nil.

Exploration and development of alternate systems, theories and scientific models continues to exist as an academic hurdle to launch over. In this way the cosmos will never give up all its secrets. Humanity will never solve all the mysteries of the universe. Great minds of the past will never be vindicated. The human race will never be taught these concepts in school. We cannot learn freely if we do not have all the necessary details required for further study. Without this, it is impossible to discern our own educated truths about creation and the purpose of our existence. Fortunately, standard model science lacks the power to prevent personal pursuit of light. It cannot stop a person from studying the concepts put forth by independent, privately funded scientists who are always busy at work turning independent thinking and alternative studies into scientific fact. Lack of acknowledgement of outside the box paradigms, models and systems does not limit our ability to learn other perspectives. There are several credible alternative scientific theories that have been presented to humanity for ages that do not align to standard model thinking. But this fact alone doesn't render any of these alternative theories incorrect. It simply indicates that the standard scientific model rules the academic world of science. It is a tightly run world that has no time or money to contribute to abstract change of any sort.

Immanuel Velikovsky made several important contributions

to Science that many average everyday people are often unaware of. He made numerous intelligent observations and revelations about the force of gravity. This work ultimately reshaped the face of gravity as we know it into a more efficient, logical working model of gravity. His theory and paradigm regarding the force of gravity quintessentially corresponds to the scientific processes and functions of universal Law operating in an electric universe. While his work relating to gravity was of very sound logic, providing an abundance of testable, observable evidence, this study was only expanded upon by a few scholars. One of these great academic minds was Wal Thornhill. As a result of intense research and extended study, Wal Thornhill provided an abundance of logical observable evidence about the force of Gravity. Very abstract teachings from the known standard model of gravity were bravely and fearlessly explored. In many cases they were also proven. But alas, to change the standard paradigm of accepted science regarding the force of gravity, the entire lot and collection of all scientific books and papers would have to be rewritten. Can you imagine the insanity that would cause in the world of modern science? It would be like a hurricane happening in a file room. Paper, pencils, pen protectors and reading glasses would be flying though the air all over the place, sticky notes would be stuck to everything in sight. Academics would have to take shifts working round the clock in order to rectify the issue. It would be outright bonkers, total madness.

In his 3rd JWST observations Thunderbolt video review, upon which this current chapter expands, Wal Thornhill reminds us of the humble brilliance of Isaac Newton. He resurfaces a wise statement Newton made in a letter he wrote to Robert Hook in 1975. In the letter he made the following statement, *"If I Have Seen Further, it is by Standing on the Shoulders of Giants. It is Essential to first find the Historic Giants Facing in the Right Direction."* This comment essentially sums it all up very nicely. An expanded awareness and objective open minds contribute to expansive evolutionary science. Close minded tunnel vision, the inability to look outside of standard models set in place is an ineffective mode of thinking. It serves only to contribute to delaying our scientific revolution and conscious evolution. To me facing in the right direction, encompasses the need to seek further light in ancient history and better guidance along the way. We do this by digging deeper into the past as this is where we find the giants in our history. Both in the animal kingdom and in kingdoms of humanity. The ancient past is also where we find the Gods of antiquity, the myths of mankind, the historical legends and heroes. Hercules, Jason and the Argonauts, Odysseus, Zeus, Apollo and so on. Within these stories and tales is where we find the clues and the truth that we all seek. There is much more truth to mythology than fiction. One way to learn this is to raise our academic gaze and look up at the stellar cosmos through the eyes of Immanuel Velikovsky, Wal Thornhill and David Talbott.

In the first book of this series, I made mention of Quasars and a process that describes a form of galaxy reproduction which technically involves reflective galactic mirroring. I offered only a brief description of a quasar and the celestial phenomenon that they emerge from and evolve into. As we continue, we will move further into the study of these luminous quasi stellar objects. It is now time to investigate Halton Arps quasar origin model of as it relates to galaxy formation. In order to determine the credibility of these findings for ourselves, further research must be conducted. It is imperative that we take a logical, rational look at the well documented, heavily evidenced science presented by astronomer Halton Arp, the Galileo of Palomar.

Regardless of this open minded alternative study of Arps work regarding redshift and quasar development, I won't say that NASA is wrong or that they don't rule. In a nutshell, NASA is epically cool and absolutely does rule. Despite it's known strict nature and unwavering unwillingness to be budged. All I could ever ask of modern science on behalf of the enlightenment of our human race as a whole, would be for standard model scientists to give just a little credibility to the brave, courageous independent free thinking minds of science. Just a wee little bit of telescope time to investigate their logical, rational discoveries would also be very nice if you would be so kind. Humanity would thank you. Ultimately it's time to commission all astronomers, astrophysicists and cosmologists to bring forth the same radiant light.

CHAPTER 8

Exploring The Origin Of Galaxies And Quasars From The Radical Expansive Perspective Of Halton Arp

Halton Arp received his bachelor's degree from Harvard College 1949. He received his PhD. from California Institute of Technology in 1953. He was one of the world's most inquisitive professional Astronomers and a brilliant Astrophysicist. Earlier on in his career he conducted Edwin Hubble's Nova search in M31. He has earned the Helen B. Warner Prize, the Newcomb Cleveland Award and the Alexander Von Humboldt Senior Scientist Award. For 28 years he was staff astronomer at Mt.Palomar and Mt.Wilson Observatories in California. While there, he produced his well known catalogue of *Peculiar Galaxies* that are disturbed or irregular in appearance. Researching photographs and spectra with big telescopes Arp thoroughly studied the images. He discovered that many pairs of quasars (quasai stellar objects) which have extremely high redshift are actually physically connected to galaxies that have a low redshift. He also deduced that this could only be the result if these galactic anomalies were close or within the same vacinity of each other. As we learn in mainstream science, quasars with very high redshift are thought to be receding from us very rapidly and as such are said to be located at a great distance away from us. You can learn more about Halton Arp and his amazing discoveries on his official website www.haltonarp.com.

As you soon will see, Halton Arps radical observations and findings, while evidenced with endless observable proof, were rejected by modern science. This outcome is of course entirely due to the fact that accepting Arps observations of high redshift quasars actually being local anomalies would have clashed with standard model teachings. In short, that of the big bang model itself and all accepted cosmology upon which it is based. In other words, embracing his very logical, intelligent, perceivable observations, would cause a fundamental re-examination of the entire foundation of everything that we are taught by modern science. Halton Arp was a brilliant scientist. Many professional scientists and his fellow academic peers do believe that his life's work should not have been so easily discarded. Halton wrote 4 books of interest on the subject of astronomy and cosmology which you can consult for further research into his rare findings and discoveries.

Catalogue of Discordant Redshift Associations,
Redshifts Cosmology and Academic Science,
The Arp Atlas of Peculiar Galaxies,
Quasars Redshift & Controversies

After Halton Arp's death, a book was written in his honor entitled *The Galileo of Palomar*. It is a compilation of various essays in his memory. This book is a celebration of Halton Arps life work in which his findings and efforts to bring new scientific evidence to the table, are vindicated by a variety of professional scientists from various academic fields. That being said, astrophysicists are strictly bound to the standard models that science has provided for further study. Despite Arps vindication and high probability of accuracy, there will be no expected abandonment of the big bang theory or the expanding universe model by mainstream science anytime in the near future.

Essentially, the evidence that Halton Arp spent his life accumulating, directly contradicted the foundational science behind the big bang and argued against the structure of an expanding universe. The book dedicated to his life's work *Galileo of Palomar* displays Arps many achievements as well as his devout commitment to science. He was greatly admired by professionals in various fields of science and amateurs alike. As to be expected, he was heavily criticized by many standard model scientists who refused to sway from the confines of known systems to look outside of the box at new theories and observations. Despite this criticism, Halton Arp maintained his loyal commitment to his pursuit of scientific insight. He was deeply dedicated to his mission to explore space until his final day. On that day, humanity lost an impeccable, intellectual, radiant, courageous explorer whose gaze was always fixed upon the luminous cosmic skies above us. His focus on the heavens is now reflected in the stars.

At this juncture, it is equally important to note that standard model scientists were particularly in strict disagreement with Halton Arps theory of intrinsic redshift. Modern science insists that intrinsic Doppler effects and Arps conclusions about quasar redshift in connection to local stellar objects and anomalies were not consistent with any known theories of physics. Consequently, his work was not supported by any experiments conducted by the mainstream scientific community. But this set back doesn't disqualify his conclusions in the minds of everyone. Arps claims of Intrinsic Redshift were based on intelligent theories and physical observations. However his findings still presented with multiple inconsistences between observable results and standard models of science. Arp offered several types of observational evidence that conflicted with what is commonly thought about quasar redshift. Once standard models are set in stone in the world of academia they become immovable.

Indestructible paradigms are closed to outside concepts, theories and abstract or radically unusual findings. Academic science as we know it is simply obligated to stick to the program they've been given.

This state of mind in regard to space exploration serves to inhibit our scientific evolution. It is very difficult for intellectual leaders such as Halton Arp, Wal Thornhill, Immanuel Velikovsky, Nikolas Tesla and David Talbott to break through barriers such as these. The presence of such a strict pre-established structure in regard to known, accepted, scientific facts, does not indicate in any way that the research conducted outside of known standard models is in any way incorrect. It's simply inconclusive as a result of an unwilling systems refusal to spend money exploring new alternative scientific concepts which they simply deem fantastic. Often ideas outside the norm are unfairly labelled pseudoscience. In truth, Halton Arps findings were ultimately never fully researched, experimented upon or further examined by the academic standard model scientists. Thus, in the minds of many intelligent scientists, professionals, academics and researchers the mystery he uncovered remains unsolved. In regard to his research on quasars, modern science uses various consistency checks to verify the nature of redshift galaxies and quasars. Arps research, his interpretation of redshift and his concept of intrinsic redshift were abandoned by the academic science community simply because it did not fit the standard model concept of redshift objects.

Arp and his team of fellow astronomers studied and discovered connections between low redshift parent galaxies and high redshift quasars within the vicinity of each other. Of course, in the standard model of redshift this is impossible. Thus, modern science argued against his findings on account of small sample sizes claiming that the selected areas chosen for sampling were bias. So, in the eyes of the academic science community Halton Arps work was unnecessarily placed in the category of unreliable or incorrect science. This decision was indeed an oversight on behalf of the modern scientific community. Especially in the minds of experts conducting valuable, intelligent, observable independent research. Even in the minds of many of the systems own scientists it was an unfair branding and misplaced label at best. Regardless of the opinion of accepted, standard model science and the criticisms he endured, Arp received equal amounts of praise and validation from many others within his scientific field of study. It is with an open, objective mind that I am examining and presenting the brilliant observations of Halton Arp and his team of fellow astronomers. Research of this nature requires the same expansive intellectual approach that we would engage to explore the heavily researched life work of Immanuel Velikovsky. In order to develop a well rounded understanding of science and spirituality combined, we must maintain a non-judgmental, expanded state of mental awareness. Keeping an objective open mind in regard to logical, rational, observable information is a key asset for any intelligent explorer. It is in this way that we develop our spiritual consciousness and increase our ability to comprehend our place in Nature.

In this book, we are researching spiritual, correlations between science and our Source, the Absolute. Naturally, theoretical science applies to this work. Creation theories are of great significance despite the fact that modern science cannot conclusively prove them. Mainstream Science has made no claims of proof of primordial creation beyond their own model of the big bang origin. In fact, even our expanding universe is in reality, only a theory. Unless Halton Arps theory of intrinsic redshift became legitimate accepted science, there will continue to be no competition for the known standard model of redshift. That is how mainstream science works. No one will ever know if Halton Arps theory is correct unless it can be experimented upon and proven accurate or inaccurate. All who loved and respected Halton Arp and his many astronomical contributions to modern science saw great significant purpose in his lifes work.

On Halton Arps official website, www.haltonarp.com you will find an article entitled *Origins of Quasars and Galaxy Clusters.* This very well known commonly studied article contains information regarding the nature of quasars and their ability to contribute to galaxy formation. Additionally, it explains how quasars aid in the formation of cluster galaxies in the same manner as their host galaxy from which they emerge. In his research Arp found numerous observable examples of smaller galaxy clusters within the vacinity of larger active galaxies that had significant associations. He discovered patterns of paired clusters equally set across large central galaxies with similar redshifts. He further discovered that galaxies found within clusters tend to be strong emitters of x-rays and radio waves. In other words, they were highly active emitters of light energy. Halton Arp researched, observed and studied many examples of big telescope images. These included those taken by Hubble. From this research he produced a lot of very credible evidence to back his claims of quasars being ejected from large local parent galaxies. Examples were provided that displayed and revealed potential ejection of quasai stellar objects from central low redshift galaxies in the direction of higher redshift clusters of smaller galaxies. In many cases these smaller cluster galaxies even showed signs of elongation along arc lines leading back to the source galaxy. A fascinating discovery indeed.

Arp noted that cluster galaxies found within the vacinity of a central galaxy actually resemble quasars. These quasars show signs of correlation and similarities to the nearby central galaxy from whence it is proposed to have been ejected. Arp persistently defended his position in regard to his belief that these similarities provided proof that the quasars were ejected from the local active, central galaxy. He presented endless samples to back his claims. In the minds of his colleagues and many other members of his own scientific community these samples were not considered bias by any measure. According to Arp and his team of fellow astronomers, evidence of quasars being ejected from AGNs (active galaxies nuclei) has been presenting itself to astronomers for over three and a half decades.

After being ejected from active parent galaxies as high redshift objects, quasars eventually evolve to lower red shift. This change can often occur as quasars break apart forming small cluster galaxies through a process called fragmentation. These cluster galaxies now at lower redshift become low luminosity galaxies within the general locality of the low redshift parent galaxy from whence they emerged by way of ejection. This is exactly what Halton Arp was proposing. Arp firmly believed these smaller cluster galaxies to retain an eternal connection to their parent galaxy. Arp assigned the title *Quasar Intrinsic Redshift* to this process of quasar evolution from higher to lower luminosity within cluster galaxies. In most cases the observed quasars do not evolve into lower redshift galaxies until they reach a certain distance away from the central galaxy that ejected them. What we are learning through our examination of Halton Arps work so far, is that high redshift quasars and cluster galaxy groups originate from large low redshift active galaxies. Did he find any proof or images to truly back these claims? Yes he did. In fact, he generously provided many of such pieces of evidence that did indeed appear to strongly support his observable theoretical conclusions.

My spiritual interest in the scientific life works of Halton Arp, directly relates to his proposed system of galaxy formation.

This process of quasar ejection and the observable evidence put forth to support it aligns to the functions of Universal Law. Repeated patterning of the creative processes of our source is evident in the manner in which galaxies come into existence. The proposed system implies organized, orderly, intelligent design. The function of central galaxies to reproduce by way of ejecting a quasar that eventually forms a new galaxy also serves as an adequate reflection of the Natural Law of Correspondence. This law is always at work conducting greater macrocosmic processes all throughout the cosmos. Essentially, what is being described here is a galaxy's ability to produce a reflective copy of itself. This process in which a galaxy reproduces itself is modeled after primordial systems of creation and universal law. The works of Halton Arp align to Natural Law whereas black holes do not. Gravitational Lensing is a known scientific process that refers to light bending around objects in space creating optical illusions. Thus, a new term, Galactic Lensing could suitably be applied to the natural reflective reproduction process of galaxy formation. It would be an appropriate term to describe the way in which great celestial manifestations contribute to ongoing creation at macrocosmic levels.

Gravitational lensing is a scientific term that describes a commonly known cosmic phenomenon. A massive celestial body like galaxy clusters or other galactic anomalies are said to cause a curve resulting in a bending of space time. This curvature causes the path of light around the celestial body to be visibly bent, as if by a lens. The galactic anomaly or celestial body that is causing the light to bend around it, is what is referred to as the lens. Strong gravitational lensing is said to result in such powerfully bent light that multiple images of the light emitting galaxy could potentially be produced by way of optical illusion. On the other hand, poor or weak gravitational lensing can cause galaxies to appear as stretched, distorted and as such more difficult to observe.

According to this standard model of science, even gravity from planets is said to effect light, facilitating our ability to observe them through high powered telescopes. Scientists are then able to discover dim or faint astronomical bodies, celestial objects, anomalies and so on. These luminous objects are taught to be the lenses themselves. Once discovered these lenses can be studied through their gravitational effects. This model is a proper description of what modern science terms gravitational lensing.

So as you can see the known scientific process of gravitational lensing describes an optical illusion created by the bending of light around an object in space. This process does not refer to the system that Halton Arp is essentially proposing. Ultimately Halton Arp is describing a process by which a quasar can potentially contribute to the creation of small cluster galaxies filled with stars. We will examine this theory, as described by Halton Arp in search of correlations between the research he produced with his team of fellow astronomers and the Natural Laws of creation. As we know, Nature has the inherant ability to manifest biological life using infinite systems and primordial models of energy transformation. These systems are busy at work, all throughout the universe at all times. These creative processes are built on a corresponding foundation of reflective mirrored reproduction from macrocosmic levels of creation all the way down to the tiniest biological lifeforms. In this way, all things we see come into existence through an immortal, primordial system of metaphysical processes governed by the standards of universal law. These creative systems of manifestation were set in place at the dawn of time. These models and paradigms are immortal metaphysical and alchemical archetypes. What we are seeking to determine through an examination of only a limited portion of the body of work produced by Halton Arp, are quintessential spiritual connections within known scientific systems of creation. As such, we will observe the functions of quasars and observe their creative ability from a spiritual and philosophical vantage point.

The refractory process of Gravitation Lensing does not serve us justice in this particular research work. When we look at evidence of galaxy formation through the eyes of Halton Arp, a new story unfolds. It has been proposed in the electric universe theory that a nucleus of a galaxy may actually consist of a pulsating plasmoid. Not a black hole. In an electric universe, the quasar is a luminous bi-product of the plasmoid nucleus. In order to fit this model into place the highly charged galactic nucleus would have to embody a natural ability to create quasars and eject them. In the ejection process the quasar moves outward and away from the center of the parent galaxy through massive electromagnetic plasma jet streams. The quasar emerges from the host galaxy at a very high redshift with similar attributes of the nearby central galaxy. Ultimately in this model the quasar would also have the ability to emit light and break down into smaller galactic materials. Arps research revealed that quasars are quasi stellar objects that eventually evolve into a cluster galaxies through a process called fragmentation.

In the world of standard model science this theory breaks critical codes. For instance, standard model science describes the center of a galaxy, not as a pulsating plasmoid but as a super massive black hole. While studies have been performed to build a substantial foundation for this theory to be set in place, it's still a theory nonetheless. In great contrast in a study of the electric universe there are no black holes and no white holes in the cosmos. There is however, an abundance of energy found in nature that facilitates a rational, testable comprehension and acceptance of the electric universe model. An astronomer named George Abell and his collaborators began cataloguing galaxy clusters over 40 years ago. At one point it was believed that the core of galaxy clusters were comprised of old E galaxy stellar objects. These galaxy clusters were assumed to be gas free and inactive. This evidence came into question as NASA developed more advanced telescope technology with far greater optical abilities. Today, x-ray surveys reveal that the cores of cluster galaxies are actually strong emitters of x-ray light. This evidence was not easily explained by standard model science. Essentially, this would be the expected result if we were to find a pulsating plasmoid nucleus at the heart of the cluster galaxy. The exact same way that we would expect to find a pulsating nucleus at the center of the parent galaxy from which the quasar was initially ejected. Can we prove any of this? Perhaps we can.

Modern science is now well aware of the fact that galaxy clusters, large central galaxies and quasars emit jet streams of energy and light. The functions of these anomalous galactic objects have been examined, researched studied and photographed by both Hubble and James Webb. Cosmic phenomenon such as host galaxies, quasars and cluster galaxies are known to be the three main sources of extragalactic x-rays of infrared light. In addition, evidence now exists to support high redshift quasars and galaxy clusters associated with much lower redshift galaxies nearby. According to Halton Arp, newborn quasars have very high redshift and this is an indicator of youth. This theory is in direct competition and contradiction to known methods by which standard model science measures redshift. In the modern academic community of accepted science, high redshift objects indicate great age and distance away from the observer. So as you can see, Halton Arp encountered this obstacle every step of his scientific career and made various attempts to disprove the standard model. Consequently, the redshift of galactic bodies as proposed by standard model science fits neatly into the model of an expanding universe.

In order to accept Halton Arps work as submissible evidence

mainstream science would have to abandon core creation theories that are the literal foundation upon which big bang cosmology is structured. Every scientist whose curiosities take them outside the box of known standard academic scientific thinking will always encounter copious obstacles getting their ideas across. Current systems are far too set in stone to allow for sway in any alternative direction. Unfortunately, this is the stark reality that all independent researchers must eventually confront. The system has been designed to operate in this way. As we know, restrictive policies such as this are not merely limited to scientific fields of study. They extend into various other professional academic fields including geology, archeology, music, to modern and holistic medicine. This restrictive limitation is a well known fact. Humanity has simply come to passively accept it. As a result, alternative avenues of study are rarely given much interest by professionals who are committed to a study of the standard models already set in place within their chosen field of study. This great obstruction has been very perplexing for numerous academic professionals for decades. It endlessly presents as an insurmountable obstacle and more often than not, a dead end to their independent research program. Without further support and financing from the mainstream scientific community, conducting further research requires a desperate need for private funding. In many ways this unavoidable drawback only contributes to limiting the evolution of expanded awareness and wisdom within the human collective.

It is imperative that we establish a proper connection between galaxies and luminous stellar objects within their local neighborhood. Only by doing this can we discover possible correlations that support the functions of macrocosmic creation and universal law operating within galaxy formation. Taking a closer look at some well known galaxies through the eyes of Halton Arp reveals new surprises. Cen A/ NGC 5128 is a giant low redshift galaxy. Arp discovered associations linking local active higher redshift galaxies and the ejection of radio plasma from the main galaxy. He found Abell clusters that were positioned along an arc line of the x-ray radio jet of an active parent galaxy moving in a northward direction. This same trajectory line is associated with emissions of radio plasma from Cen A/ NGC 5128. So in this example there appears to be a direct astronomical connection between the main active galaxy and objects found within its direct locality, such as galaxy clusters and active quasars. There exists multiple images of this galaxy and the conditions being described by Arp after careful observation and intense analysis. In fact the many supporters of Arps work are well aware of the fact that volumes of evidence exist to support the concept of direct correlations between small galaxy clusters located within the vacinity of their giant parent galaxy. Modern Science has long since verified the fact that galaxies emit x-ray light and radio waves in the form of jet streams. These jet streams usually extend outwards in both directions from the center of the galaxy. They create massive two sided jets of powerful energetic light that can be witnessed and observed by astronomers and cosmologists. These jets can also be one sided.

In order for this effect to be possible, there must be a quasar or electrically charged plasmoid nucleus emitting energy from the heart of the galaxy. That being said, standard models of science describe the center of a galaxy as a super massive black hole that is ultimately very tiny in size. The black hole is said to produce a highly active quasar when surrounding energy falls back or collapses into the black hole. Even in this case we have the existence of an active quasar at the center of a galaxy. The quasar in both cases is positioned in the heart as a powerful, luminous object that emits light energy. Consequently, the standard model explains it very differently than Halton Arps version. Arp additionally discovered that cluster galaxies also have quasars within their construct and as such, also emit x-rays in the form of jet streams in the same manner as larger disc galaxies. Furthermore he complied evidence that revealed the potential for pairing high redshift quasars with higher redshift objects across the central active galaxy. In combination, all of these cosmic bodies showed evidence of direct association with each other. The central galaxy itself is often of a lower redshift. Arp explained this effect through a concept that he proposed called Intrinsic Redshift.

Haltons definition of *Intrinsic Redshift* explained the apparent link between small young galaxies, quasars and large older galaxies. Arp noted many examples where these visible connections can be seen. In the same central active galaxies this energy even appeared to resemble an umbilical cord of energy connecting smaller cluster galaxies to their host parent. How incredible is it to consider such a profound discovery a reality? Does this proposed model correlate to Natural Law as we know it? Absolutely. In contrast, modern science uses redshift to measure distance. Thus, quasars with high redshift are said to be at very great distances away from us. The process of continuously moving farther away is a result of the conditions created by the expanding universe model. In Arps research, redshift was a factor of age, not distance. When using redshift as a measure of distance we end up with quasars that are too far away. This creates a perceivable conflict of interest. Arps examples of galaxies interacting with active quasars appear to show them in the local neighborhood and not extremely distant at all. Definitely not millions of light years further away in deep space. In Arps work, the theory of an expanding universe does not agree with his logical, rational, observable findings. This detail resulted in his abandonment of the expanding universe model, upon which the big bang is built in order to prove his findings correct. This bold move placed him right in the center of total madness.

Herein lies the root of the tension he received from the modern scientific community in regard to their lack of interest in further explorations of the theories he was presenting. In truth, he hit a scientific brick wall. Regardless of this set back, Halton Arp continued his academic pursuits with dedicated commitment to his life's work for his entire career. At no point in time did the standard model scientific community choose to take his findings seriously. Yet, there are countless professionals who can actually see what Halton Arp was looking at in these images of the deep cosmos. As a result, a league of faithful believers from the next generation will likely further expand on his work. In a sense, the book that tributes his career, *The Galileo of Palomar* is proof enough that this is already happening. Halton Arps courageous contributions to science and his devout willingness to defend his theories in the face of harsh judgement and controversy makes this radiant trail blazer of science a total legend. As much of a legend as Tesla and Velikovsky who also remained loyal to their own ethics and scientific beliefs, unmoved from their positions. So many brilliant minds have gone under appreciated, unrecognized and unacknowledged over the course of time. There is very little we can do about this, aside from keeping an open mind while learning and appreciating who they were. A very expanded, objective, non-judgmental mindset in regard to rational information is a requirement for those who seek to attain proper knowledge and wisdom. In any case, a healthy system of beliefs is structured first upon good information that is accurate, not misleading or false. It is also built upon that which is real, legitimate, plausible, logical and rational to each of us as we perceive the world around us. Halton Arp did indeed bring excellent, highly credible information to the table of science. Will his findings ever be embraced by standard model scientists and taught in public schools? It's not very likely to occur in the next few decades, but we can dream all we wish. In order to achieve this victory, the system would have to broaden its horizon, expand its awareness, increase its perspective, welcome new logic and consider alternative evidence. Greater flexibility would be required in order to achieve a unanimous consensus amongst the world's most brilliant scientific minds. Successfully deployed, this cause could result in the effect of raising the standards of modern science to a whole new level. It could initiate and facilitate the emergence of a scientific revolution. Ultimately, as James Webb continues its meticulous search of our universe, a new scientific frontier is breaching the waves of the cosmic sea. This result is inevitable with advanced technology such as this at our disposal.

Now that we have a basic understanding of what Halton Arps body of work entails, it is time to look for some spiritual correspondences. The standard model scientific community dismissed Arps findings as quickly as he abandoned their theory of an expanding universe and their model of redshift. If only these great powers and intellectual minds could have come to a compromise. Space exploration and further advancement of humanity's scientific knowledge is a worthy cause for an academic truce. Ultimately we would all be illuminated and enlightened in effect. Examining Arps body of work paints an entirely new image of celestial creation and cosmic evolution. We are learning how galaxies may very well have the ability to create identical mirror reflections of themselves. This proposed process of reproduction aligns to known metaphysical systems of primordial creation and as such, also to Universal Law. Halton Arp and his team of fellow professional astronomers reached the following final conclusions. They determined that quasars are ejected from the nucleus of active host galaxies. New born quasars have very high redshift. Eventually, quasars evolve into lower redshift galaxies. This evolution often takes place when a quasar reaches a certain distance away from its parent galaxy of origin. This reflective process of reproduction is infinitely conducted through the orderly scientific forces of Nature. It is significant that we examine these phenomenon in search of quintessential connections to the functions of Universal Law. Do Arps findings compare to processes that are occurring in our microcosmic biological world at all times, all around us? Yes they do. This is the essence of the Law of Correspondence at work in creation.

Astronomer Halton Arp, the Galileo of Palomar,
a shining star. March 21 1927-Dec 28, 2013

CHAPTER 9

Objectively & Subjectively Reviewing The Observable Evidence Presented By Astronomer Halton Arp The Galilaeo Of Palomar

The following chapter is a objective and subjective review of Halton Arps research into the origin of Quasars and their association with potential host galaxies. We will be looking for the quintessential connections between Arps findings and the scientific process of Nature that are governed by Universal Law. For further research into this study, see article by Halton Arp, *Origin of Quasars and Galaxy Clusters.* This article can be found on Halton Arps official website, www.haltonarp.com .

Insipient quasars are newborn quasars in the initial stages of development. Halton Arps research shows us that x-ray sources ejected from active galaxies can sometimes be born as pairs or triplets. This increases the possibility of the quasars being *insipient* which refers to the process of fragmenting. When these high redshift quasars break down into smaller pieces through fragmentation they evolve into groups and clusters of lower redshift galaxies. In Arps research, this is how luminous galaxies are born. This process describes a system in which quasars are ejected from a host galaxy indicating the reflective nature of primordial creation patterns already present in the universe. The scientific processes of Natural Law govern and regulate all systems found within Nature. These laws are in effect from macrocosmic to microcosmic levels of creation at all times. The geometric and fractal patterns of creation are immortal constructs that are infinitely repetitive all throughout our biological world. The same set of laws apply to all things in existence. That being said, there are tremendous amounts of correlations to universal law at work in Halton Arps model of quasar galaxy interaction. These connections will be further explained as we continue our review of Arps body of supporting evidence.

First, let's shift our focus to a couple of cosmic samples along with Arps explanation and analysis of these galaxy images. To support Arps claims of quasar duality, we can find a number of observable cases where pairs of quasars or cluster galaxies can be found within the cosmic neighborhood of a large low redshift galaxy. In his book *Atlas of Strange and Peculiar Galaxies,* Arp shows us multiple examples of this phenomenon. These findings lend powerful credibility to his claims. As abundance of observable evidence can easily be seen by the naked eye when examining these high powered telescopes images. Especially those of Hubble and the James Webb Space Telescope. As we inspect the images in Arps book of galaxies, it's difficult to logically deny what appears to be so clearly obvious. In the case of standard model scientists, they have no other choice but to reject the data. This is based solely on their commitment to standard models of science already well established in our academic world. Sadly as a result they have yet to explore his ideas and concepts in the proper laboratory atmosphere using the best equipment and latest technology. To put it to rest once and for all, we would need to set the most intelligent scientific minds of our modern era to the task. It is critically important that issues such as these be addressed and resolved. If we wish for a scientific revolution that ushers in the age of light we must all expand our awareness and consciousness levels, eliminate judgement and adopt intellectual flexibility.

Further examination of examples provided by Arp and his fellow astrology colleagues provide evidence indicating that quasars are always found close to x-ray jet galaxies. There is an observable patterning effect. It can be seen in multiple cases. There is a very low chance of this being accidental. Especially in consideration of the fact that all of creation is of intelligent design and that there are no accidents or chance coincidences in metaphysics and alchemy. Sacred geometry and the platonic solids are globally accepted models and systems of metaphysical creation. These forces are already proven by sciences such as sacred geometry, bio-geometry and quantum biology. Those who study these sacred sciences are already well aware of this fact. Modern science does not deny isotropy, consistency of patterns and biological lifeforms found all throughout nature. In many of the cases he researched, Arp discovered that smaller cluster galaxies surrounding a giant host galaxy seemed to routinely follow the same or similar arc line leading back to the initial source of ejection. The origin point in this case was the active nucleus of the local neighborhood parent galaxy. This appears in many of his image samples. He provided endless examples. The cases clearly showed these co-linear arc lines along with strings of quasars and galaxy clusters that could only be emerging from these giant local galaxies. In actuality, the evidence he presented was abundant enough to win his case hands down.

Essentially Halton Arp was attempting to prove exactly what I am looking for as a seeker of light and truth. The quintessential connections between the scientific processes of Nature, our spiritual existence and Universal Law. He was desperately trying to prove to modern science that the pattern of galaxy formation he was examining was a natural repetitive pattern found in nature and in no way merely chance or coincidence. Halton Arp committed much of his life to a diligent luminous hunt for obvious, evident existence of connections between quasars, galaxy clusters and host parent galaxies. He produced volumes of radiant evidence. He brought a cornucopia of intelligent observations pertaining to natural galaxy formation into the scientific arena. He presented more than an adequate volume of evidence to prove that his study deserved credibility. The many observable cases of higher redshift quasars being ejected from larger galaxies demands further examination by the modern scientific mainstream academic community. Regardless of whether or not his findings disagree with the standard models of science. Whether or not standard models are reinforced or proven right or wrong it is a challenge worth facing. This would certainly put an end to the mystery.

Arps body of work shows that galaxy clusters are often aligned with the parent galaxy as an extension of their host. His examples show that these smaller bodies form chains, sub-clusters and filaments of luminous celestial objects leading back to the source galaxy. These sub-clusters and so called independent anomalies have similar redshift values. Some of these cosmic stellar bodies can clearly be seen to form an arching line moving outwards from both sides of the nucleus of the central galaxy. The jet stream of light and energy that emerges from quasars and large galaxies always originates from the nucleus at the center. These arching lines show elongation on a proper path leading away from the central galaxy, the place of origin from whence they are rationally presumed to have been ejected. This effect is something that any intelligent person can see for themselves upon simple examination of the images in Arps *Atlas of Strange and Peculiar Galaxies.* It doesn't take a scientist or an astrophysicist to observe what Halton Arp was looking at. However, in order to embrace this theory we have to let go of the concept of redshift being measured as a value of distance. In addition, we must set aside the model of an expanding universe, temporarily. That paradigm simply does not apply to Arps rational, alternative perspective on the nature of quasars and galaxy formation. The evidence he shows us, thus defies the known laws of standard model science. That being said, science must some day face the inevitable and merge with the unexpected. In our modern age of light, we may likely witness this immanent change. Eventually.

Modern astronomers have truly been given great purpose for conducting further research of Haltons findings, theories and concepts. There is much evidence in existence to back his case. Of course a study of this nature could serve to rattle the foundation of known accepted standard models of science. On the same token, it would also provide substantial ground for new emergent theories to be validated and researched in the modern scientific community. Further study, experiment and research could then be conducted. This is how new discoveries become fact. One thing that science is about is courage. It takes a great significant amount of courage to hop into a turbo jet powered metal cylinder and blast off into space. I certainly don't have the courage to do it. Likewise, it takes a tremendous amount of guts for any professional scientist to take a stand against standard model science. Halton Arp, had plenty of both. His courage and guts never quite earned him the glory he sought despite his powerful case.

Galaxy M101 is located to the south west side of famous radio galaxy 3C295. These two galaxies are massive celestial bodies with remarkably similar qualities. A study was conducted of the very bright quasars found within the atmosphere around Galaxy M101. Both 3C295 and M101 have quasars that are distributed along the same general arching line of smaller cluster galaxies. These bright, high redshift quasars are located near each of the galaxy clusters. One could easily logically assume that these are newborn quasars that may be connected more so to the smaller galaxy clusters than to the central galaxy nearby. If they were not ejected from M101 itself, then the possibility arises of these quasars being ejected from the nucleus of the galaxy clusters. The most active cluster galaxies and quasars surrounding Central Galaxy M101 are positioned in a line extending outwards from the center of the central parent galaxy in a NW-SE direction. This shows us the assumed dual jet stream ejection path apparent in known quasar functions. Most of the local active galaxies near M101 are clearly seen to be positioned exactly in this way. This logically promotes signs of direct association between these galactic phenomenon and cosmic bodies.

In order for standard models of science to work efficiently with Arps study of quasars the standard model, of redshift would require adaptation. Standard model science places its stock on determining the distance of quasars based on redshift. Their work must reflect and be structured upon the currently known, already accepted redshift paradigm. In the standard model of redshift, objects with incredibly high redshift are said to be at a greater distance away from us than all other bodies with lower redshift. This is a contradiction to what can clearly be optically perceived in the case of Galaxy M101. There are also quasars located at greater distances from the main body of M101. It is quite rationally possible that these quasars may have originated from the nucleus of secondary bodies. The source of ejection would then point directly to the smaller Makarian cluster galaxies that are surrounding M101. This is exactly what the image appears to indicate as well as what Halton Arp is clearly showing us and attempting to validate.

The brightest and earliest discovered 3C Radio sources reside in the neighborhood of Galaxy M101. They are mostly located along the NW-SE arc line of stellar objects passing through the heart of M101. Galaxy 3C277.1 shows association with MRK 231 and with Galaxy 3C295 nearby. All of these celestial bodies show associations with M101 itself. Some of the other cluster galaxies seen in the general locality may be associated with other nearby galaxies such as MRK 477. Arp concluded that it is logical to assume the possibility of a *Cone of Objects* being ejected from Galaxy M101 in both directions of the x-ray jet stream on the NW-SE axis. Arp also suggests that this configuration of celestial objects could also be interpreted as two separate narrow path lines of ejection. One at an angle of P.A=110° and the other at an angle of P.A=147° from the central positon of the main host galaxy M101. The following images of diagrams produced by Halton Arp explain these conclusions and the close proximity of the luminous objects surrounding M101. The arc line of active bodies can evidently be seen on the described NW-SE axis. Images to follow show the highly active radio galaxy 3C295. A much closer image of this amazing galactic anomaly and the visible connections between the quasar and the active nucleus of the galaxy can be seen in Arps *Atlas of Strange and Peculiar Galaxies.* To follow, are images of the highly active radio galaxy 3C295 as seen at a great distance along with diagrams of M101 and surrounding celestial phenomenon.

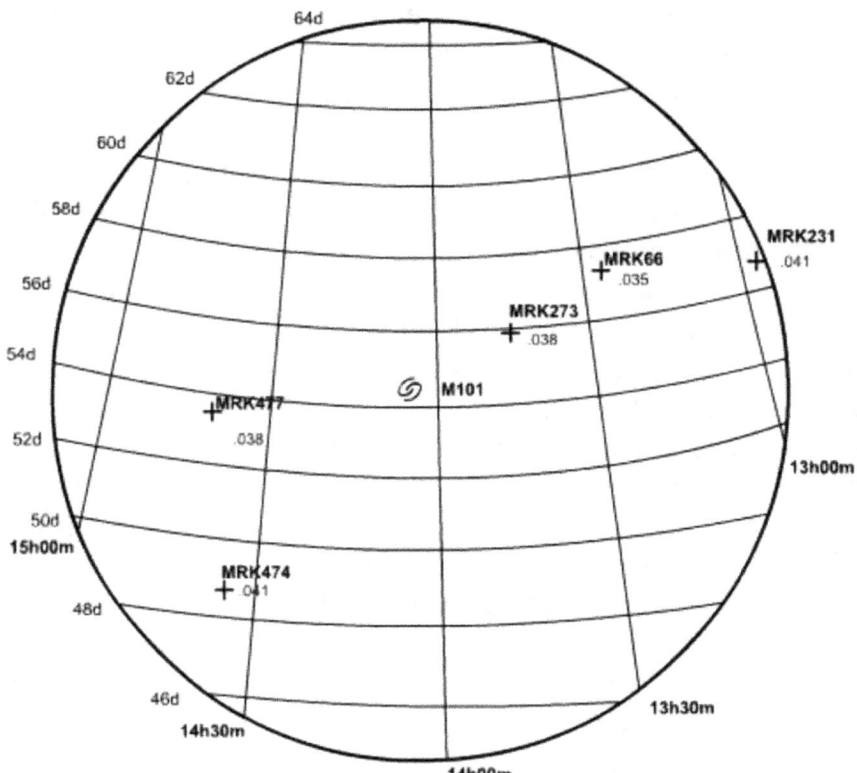

Galaxy M101, showing nearby, luminous bodies and local quasars. See article by Halton Arp *Origin of Quasars & Galaxy Clusters*, www.haltonarp.com

X-Ray image of Galaxy 3C295, NASA www.nasa.gov

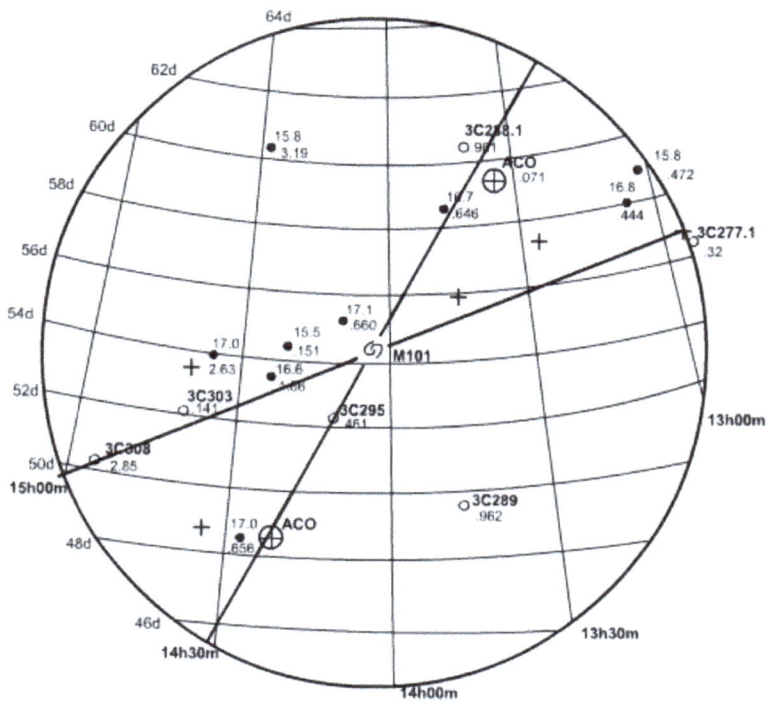

Galaxy M101 showing adjacent luminous objects and cluster galaxies on the NW-SE axis arc line. Article, Halton Arp, *Origin of Quasars & Galaxy Clusters* www.haltonarp.com

X-Rays from a Galaxy 3C295, NASA www.nasa.gov

Famous radio galaxy 3C295 shown in these two images is situated to the SW of Galaxy M101. In the 1960s, 3C295 was granted the status of the brightest known galaxy. It's redshift measured at Z=0.46. Galaxy 3C295 maintained this illuminating title until 1975. As technology has significantly advanced since the 1960s, our ability to observe and view stellar objects and celestial bodies has increased along with it. Recent observations of 3C295 by Chandra show a pair of x-ray condensations coming out of the nucleus at a position of P.A=144°. The image above shows the Galaxy along with its quasars from a great distance. To find energy ejecting from the nucleus of a galaxy at an angle of 144° is of tremendous significance to me. It is a correlation to universal law on a macrocosmic level. Perhaps it's just a coincidence, however its one that is duly noted on my part. Fundamentally, 144 is a sacred geometric number in metaphysics. It represents the total number of points on the Isotropic Vector Matrix. This sacred geometric archetype was discussed in the first book of this series *Macrocosmic Reflections in a Microcosmic Mirror.* This angle and alignment within the radio lobes of 3C295 at 144° degrees is an interesting correspondence to a sacred number regardless of whether it's a coincidence or not. The ejection origin or radio lobes is known accepted academic science in the mainstream scientific community. It is also a well known fact that x-ray jets are found at their core. In our electric universe, where we find a highly charged electromagnetic nucleus we find a pulsating plasmoid. This is the anomaly that will be at the heart of the matter. A plasmoid emits energy and light. After intense study and thorough inquisitive analysis, Arp and his team concluded that the x-ray sources in 3C295 are clearly indicated to be *in the process of ejection.* This radiant Galaxy is surrounded by other large galaxies, smaller cluster galaxies and quasars. In my personal research this scenario is exactly the set of circumstances that I'm looking for. When evidence of this nature presents itself, I pay attention. To neglect or ignore this knowledge could interfere with my mission of academic, spiritual development.

To deny the credibility of discoveries such as Halton Arps could hinder the scientific evolution of our race. If Arps findings are accurate, what we are witnessing is the art of creation in action within the heart of Galaxy 3C295. Modern science now has even more amazing tools like the James Webb Space Telescope. As we are well aware, this highly advanced technological masterpiece has the ability to glare more deeply into luminous cosmic phenomenon existing in deep space. As such, humanity is at great advantage. Modern science is now capable of exploring space much more intimately in its pursuit to reveal the secrets of our universe. Studying images of 3C295 shows us massive systems of creation in progress. In this light humanity can examine these images which ultimately expose creative models governed by Universal Law in an active state of macrocosmic operation. It is a Ritcheous, spiritual privilege to behold the creative systems of the Absolute in action, even from a distance.

Are there other environmental habitats in the universe suitable for earthlike planets and biological lifeforms to exist, grow and thrive? Science says, yes. There is a lot of highly charged activity in the local galactic neighborhood of Galaxies 3C295 and M101. These galaxies and their nearby surrounding quasars share similar characteristics that connect them together like stellar family units. Between 3C295 at a redshift value of Z=0.461 and it's closest neighbor at Z=0.46, we find another galaxy at Z=0.285. The two younger higher redshift bodies appear to be almost perfectly aligned in an arching line across the central lower redshift galaxy. The line is extending at 11 arc seconds on either side of this central galaxy. To a practioner of metaphysics this is clearly a divine reflective pattern of a scientific process of creation repeating itself within nature on a macrocosmic level. There is only a very slight chance of this copious amount of galactic activity of similar astronomical function and structure all occurring within the exact same cosmic neighborhood. Halton Arp insisted this, telling us that this was in no way accidental. There is a much stronger and more likely chance that all of these celestial bodies and galactic phenomenon are intricately connected. By way of astronomical origin and divine intelligent design all cosmic systems originate from one ultimate source. This divine reproduction system is governed by the forces of Universal Law. From an expanded esoteric and metaphysical perspective this is exactly the case.

Arps proposal of galaxy formation is a prime quintessential example of the connectivity between scientific, archetypal patterns of creation and Natural Law. The law of Correspondence tells us that the primordial origin model of creation reflects all other scientific processes, models and systems that repeat themselves all throughout nature. In quantum physics and eastern philosophy this effect occurs quite naturally by way of intelligent design. In this book we are exploring Halton Arps work in regard to a galaxy's ability to reproduce itself through a process of natural self replication. This function is shared by the parent galaxy, it's smaller cluster galaxy siblings and evolving infant quasars. This system does not rule out the standard model of gravitational lensing. That particular process describes an optical illusion that is created by refraction in which light which is being bent around objects in space. This cosmic phenomenon has been cited as an argument against the evidence presented by Halton Arp and his fellow colleagues. When looking at these two independent systems separately as two distinct scientific models and processes there is very little connection between these two phenomenon in this particular case.

The image of galaxy NGC7319 shown in chapter 4, taken by James Webb in the infrared offers greater optical clarity and much more insight into this type of cosmic phenomenon. Even in Hubble images what can be seen when the photos are expanded appears to more strongly support Halton Arps theories over those of standard model science. Mainstream science still insists that we were seeing an optical illusion caused by gravitational lensing. In many ways Arps evidence equates to well researched, submissable scientific material that seems to reasonably qualify and support the logical, intelligent claims that he made. Arp stood his ground in the face of adversary and never gave up on his dream of giving the world accurate science as he saw it. He kept his dignity and firm loyalty to his beliefs and causes throughout his entire academic career.

To recap what we have learned, Arp proposed that quasars are born by way of ejection from the nucleus of active central galaxies. Both the galaxy and the quasar can emit light in the form of x-rays and microwaves. Quasars can break down into materials that form cluster galaxies. Quasars can also eject galactic material in the same way that a larger galaxy can. Ejection occurs from large luminous jet streams that extend from both sides of the nucleus of these cosmic bodies. These plasma energy jets can also be single sided. As part of their evolution, quasars eventually break down into pieces through a process of fragmentation which forms smaller cluster galaxies. This system reflects the governance of the Law of Correspondence over all aspects of the reflective nature of macrocosmic creation. The Law of Correspondence clearly teaches us that, what is above us in macrocosmic creation, is also reflected in all forms of microcosmic creation below. In other words, microcosmic manifestations will always be a reflection of the macrocosmic origin model. This law is in effect and is operative within all levels of creation great and small at all times. In metaphysics, we already know that this quintessential information is accurate and correct. This has been known for thousands of years by both ancient and modern mystics. In a world where quantum physics, geometric archetypes, sacred Fractals and other primordial patterns of creation are known sciences, this should be no surprise.

We and our earth exist in a massive galaxy that reside in our great, colossal universe. Our milky way galaxy is an astronomical model that is only one out of many galaxies that exist in the cosmos. In the standard scientific model, the milky way galaxy is said to be 480 billion years old. Regardless of its age, within this massive flat disc galaxy resides our earth which represents our whole known world. It is actually very tiny. Within the great magnitude of the expansive stellar structure and dusty composition of the Milky Way, our earthly abode is little more than a spec of space dust. Our entire human race is even tinier. It is super microscopic by comparison. To think we are all alone in the cosmos is a momentary lapse of reason strongly leaning on the side of egotistical vanity. Imagine that. When logic is applied to the odds, we can almost safely assume the high unlikelihood that we are the only biological life forms of our kind in existence. Especially when the colossal size of our universe and the functions of Natural Law are taken into reasonable consideration. Earth like planets and different versions of humanity can very easily even exist right here in our own local galaxy. In fact, once we develop an understanding of the scientific processes of Nature we can easily see the high probability of this. Universal Law teaches us that the same patterns of manifestation repeat themselves all throughout creation. These primordial systems are immortal, omnipresent and governed by the same laws that rule all aspects of Nature. The theories, findings and discoveries brought forth by Halton Arp explain the natural creative systems operating within the cosmos. These automatic systems integrate and co-operate quite easily with theories and models of both the electric universe and the functions of Universal Law. Ultimately from a metaphysical perspective, this lends much spiritual credibility to his scientific discoveries.

Galaxies can eject light via one jet or two jets, one on each side of the active Nucleus. Radio Galaxy 3C303 only has a one sided jet. For this reason, it is of particular interest to astronomers and is often studied. This is a tight, compact radio galaxy of Z=.141with its jet stream pointing outwards toward a double radio lobe. The northern section of its radio lobe is at 280°. It is within the vacinity of M101 which is at P.A=289°. In the 70's, astronomer Margaret Burbidge conducted a study of three ultraviolet objects apparently associated with the western lobe of Galaxy 3C303. Testing the spectrum of one of these luminous objects rendered a redshift of Z=1.57. Regretfully this information as well as the experiment itself then came to a halt as no further research or telescope time was assigned to her project. As such, the other two local near distance quasars were never tested for redshift. Any existing correlations could not be discovered and catalogued. Regardless, this evidence presented credible proof of local quasars within close proximity to larger host galaxies. This evidence also defied the standard model of redshift by clearly displaying that high redshift quasars can be closer than predicted by mainstream science. In the standard model, high redshift quasars exist exclusively at vast, great distances away from the other lower redshift bodies being reviewed. Based on the redshift measurements which were recorded in this experiment, the results and findings conflicted with and defied the known, accepted laws of Science. In light of this, the study was abruptly concluded and abandoned. Further research into the three quasars located within the radio lobe of Galaxy 3C303 may potentially validate the inquisitive, perceptive efforts of Margaret Burbidge. If JWST looks deeper into this matter, it could possibly be resolved quite effortlessly.

As we know, Arp proposed the concept of *Intrinsic Redshift*. What exactly does this term imply? Arps model of Intrinsic Redshift attempted to logically explain cosmic phenomenon that the standard model redshift could not. Intrinsic redshift explained how high redshift could rationally be assigned to local quasars. He explained that quasars are younger galaxies that are potentially linked to older galaxies. In some cases they even presented with visible, observable connections. In a few special cases these links optically showed energetic structures similar to the shape of an umbilical cord. Evidently, these energy X-ray condensation lines appeared to connect the quasar or smaller cluster galaxy to its parent galaxy. Standard model science uses redshift as a measure to indicate the age and distance of a luminous object. In the accepted, well known standard model the higher the redshift, the older the quasar and the greater the distance the quasar is away from us. This conflicts with the obvious, observable connection that ultra luminous quasars have with nearby local galaxies of lower redshift. If they are as distant as they are said to be based on the standard model of redshift, what we can see with our own eyes in the evidence presented by Halton Arp contradicts their paradigm. Consequently, his model of intrinsic redshift was harshly rejected. Scientists must never outright defy the ruling laws of modern science if they ever seek to receive funding or credibility for any work done in their scientific field. In Arps case this point certainly was driven home very strongly but as always, he stuck to his resilient standards and forged onwards.

To further explain his concept of intrinsic redshift Arp proposed that the matter found within quasars is less massive in comparison to other matter dispersed throughout our solar system. As a result, the hydrogen within the chemical composition of the quasar is lighter than hydrogen found on earth. Fundamentally, the hydrogen within the body of the quasar is redshifted in accordance to the light spectrum of hydrogen on earth. In this way, the high redshift of a quasar does not indicate distance. Ultimately the high redshift would indicate youthful age. As we know, all matter starts out with low mass. Eventually mass condenses into biological matter. The matter gradually gains more weight as it transforms and evolves into some form of biological life. These functions are similar to the metaphysical process occurring within the tesseract heart of Metatron's Cube. It's vocation is to crystalize light energy into matter. The plasma light energy gains weight and form as it transmutes and transforms into biological material life. In Arps theory of intrinsic redshift, light from newborn matter is red while much heavier, local adjacent counterparts such as nearby central host galaxies will have lower redshift.

In order to validate Halton Arps conclusions and theories regarding his proposed concept of intrinsic redshift, the modern scientific community would have to look much deeper into this abstract model. As observed by the leading astronomers at Mt.Palomar observatory, Arps intrinsic redshift model could easily be validated by comparing the light spectrum of hydrogen to that of its heavier isotope deuterium. Examining this evidence further would certainly be worth the minor effort and time it would require of busy standard model scientists. Especially in the event where the system is also given the opportunity to further prove, validate and vindicate its own models in order to lay all false claims to rest. It's a worthy scientific challenge that Halton Arp was always open to and more than willing to accept.

The standard model of accepted science uses the Doppler Effect to measure the distance of objects in space. Thus, in the minds of professional modern scientists this method of associating redshift and distance fully qualifies and can be verified through experiment based on standard model relativity. Case is closed. Door is shut. On the other hand in regard to standard models of redshift, *Intrinsic* Doppler shifts as suggested by Halton Arp, simply do not qualify as they do not correspond to the status quo. Arps theory of Intrinsic Redshift remains inconsistent with any known accepted theories of physics. The existence of Intrinsic Redshift proposed by Arp was never experimented upon by any members of the professional mainstream scientific community. As a result, the existence of this phenomenon remains unrecognized and unsupported. His very intelligent, highly credible claims along with his bottomless pit of supporting evidence regretfully got tossed to the way side by bigger business. Accepted Science is set in a foundation of stone. Credible findings such as this could potentially revolutionize what we know about the true nature of quasars and their connection to local, active central galaxies.

Consequently, in great contrast these discoveries were simply rejected and became categorized as inconsistencies between observation and known, accepted cosmological models. This fate is regularly met, despite how much new evidence may qualify or how obvious or legitimate the proof appears to be. Again, alternative concepts, outside the box thinking and radical new models or paradigms are not easily embraced in any field of study that already has its own pre-established system of belief. Many professionals from various fields of academic study and practice, all around the world, meet obstacles such as this on a regular basis. If we wish enlighten our human race we must come to recognize barriers such as this as a set back for rational, intellectual, expansive, inquisitive, open minds everywhere.

This is a devastating consequence and effect to face for professionals who are whole heartedly engrossed in and committed to their academic causes. Somehow, this malady must find a way to remedy itself. Great legendary, philosophical minds are defined by their vast, insatiable, open minded, non judgmental pursuit of truth, light, knowledge and wisdom. All good information is to be considered. Sources are not to be judged unless deemed unworthy, lacking in logic, rationality, intellect or are void of truth. Feasible, credible scientific evidence should never be discarded until it is further experimented upon by standard model supporters. This work should always be conducted by modern science in order to conclusively prove or disprove any new valuable evidence to be correct or incorrect. This is how science should work. Yet alas, we have a long way to go before standard models of science and alternative scientific theories can share a legitimate, common platform in any professional field of scientific study. Human minds are capable of expanding awareness and consciousness wide enough to embrace the fact that all things in our physical plane are *One.* All of creation, everything in nature is connected by way of intelligent, spiritual, metaphysical design. It is imperative that we make the decision to drop the illusion of division if we ever wish to exist in a state of profound unity and *oneness* with each other and nature itself. Only in an elevated mindset such as this can we even hope to achieve a successful union with inner self or divine alignment with the vast collective consciousness of our Source.

CHAPTER 10

Connecting The Quintessential Dots Between Arps Concept Of Galaxy Formation, Metaphysical Creation Models & Universal Law

Halton Arp proposed the high likelihood of quasars being emitted in pairs and sometimes triplets. This information is of significant spiritual interest. The first impression that comes to mind is the force of duality. This force is represented by the first movement of the seed of life. In this oscillation, the Dyad is formed creating the Vesica Pisces from which a massive stream of light energy comes forth. Like the proposed nucleus of the quasar and a host galaxy both of these celestial objects emit the light of our Source, as does the Vesica Pisces. The dyad is the sacred archetype that represents the force of duality. In metaphysics this divine symbol of creation forms during the 7 days of Genesis in the 1st oscillation of the alchemical construction of the Seed of Life. We have already learned that evidence presented by Halton Arp indicates a quintessential pattern wherein smaller cluster galaxies originate from local parent galaxies. What is actually occurring is the self replication of an identical reflective model of an energy form from a host. In a sense, what is being created is a micro model that is akin to a reflection in a mirror.

Arp devoutly believed that his evidence and findings were accurate, legitimate and sound. He showed us such obvious correlations between active X-ray emitting galaxies that it's logically difficult to easily dismiss or reject his findings. Many independent professional scientists feel that his discoveries adequately provided one piece of evidence after another to back the validity of his claims. Arps findings appear to optically indicate the impossibility of such close connections being random, accidental, coincidental or chance. As explorers, each of us are left to determine for ourselves what we see in the image samples he provides in his book of *Peculiar Galaxies* and in the many scientific papers that he authored. Read his books, study his findings compare Arps conclusions to those of mainstream science respectively. It is in this way, that you form your own belief system based on your own internal sense of reasonable, good judgement. As biological lifeforms with rational, cognitive thought capabilities, we are equipped with a vast mental capacity and free will faculties. These assets are precious gifts from our Source. Each of us is expected to use our human mental faculties to examine all information available in order to develop our own concept of truth. It is our obligated human responsibility to acquire knowledge if we wish to evolve as a species. We are here to learn and to experience the gift of Nature so generously offered to us by our Creator. We are here to experience the unconditional love upon which all of creation is foundationally built and governed by Natural Law. All macrocosmic systems of creation repeat themselves all throughout Nature right down to most miniscule lifeforms.

The way that Halton Arp describes a Galaxy's ability to replicate itself falls directly into the constitution of Universal Law. In human reproduction, a mother gives birth to a child. The child naturally never strays far from its parents until they reach a certain age. Then by way of nature, they move away from their parents and in most cases, have children of their own. No matter how far a child strays from home, they will always maintain the character traits they embody which are similar to those of their parents. We all know how this family unit forms down here on earth at our microcosmic level of existence. It is quite evident that this glorious pattern of reproduction is nothing new. We are seeing a biological reflection of a massive cosmological reproductive process already occurring at mega sized macrocosmic levels of creation. In this light, human reproduction mirrors that of galaxy formation and stellar family units. From this mystical perspective one can truly acknowledge and become spiritually aware of how everything in existence is connected. These magnificent correlations fall within the structure of the Law of Correspondence. As Above, So Below.

For the sake of attempting to bridge a gap between science and spirituality we are looking more closely at astronomer Halton Arps work in search of quintessential connections. All thorough scientific research must also take into consideration the vast pool of evidence that strongly supports the electric universe model put forth by experts in various fields of scientific study. Bearing these teachings and scientific philosophies in mind, we reach a whole new level of understanding and gain further insights into the scientific processes of Nature. In this way, we also develop a greater understanding of our Source. It is imperative for humanity to acknowledge the divine correlations that connect our existence to the spirit forces of Nature. Reaching unified levels of awareness such as this contributes to spiritual awakenings of tremendous magnitude. In the electric universe teachings, galaxies have a highly charged pulsating nucleus at its divine center. In contrast to mainstream science teachings, there is no black hole. In the electric universe model, black holes are not necessary and do not exist. Their proposed scientific functions are not required. For a moment, let us presume that the heavily evidenced electric universe theory is indeed correct. From this vantage point Halton Arps work is in total alignment with Nature. The highly active energetic electric nucleus of a galaxy ejects new born quasars. The quasar itself, also has an active electric nucleus that contributes to even further development of small cluster galaxies. Both of these cosmic anomalies emit vast amounts of radiant x-ray light and galactic material outwards. All quasars eventually fragment into smaller, highly charged galactic anomalies. The functions of this process are all electrical by way of construct.

From Arps perspective, the newborn quasar can be viewed as the infant child of the mother galaxy. As offspring of this galaxy, the quasar would naturally bear striking similarities to the parent galaxy by way of function and chemical composition. We humans also inherit the blueprint of character traits and other biological markers of our parents. These genetic codes become part of our system programming through blood and DNA. During its evolution, a quasar breaks down into multiple small units through a process of fragmentation. Its redshift eventually decreases with growth, maturity and age. This explanation shows the patterning that we are looking for in Nature. A system of galactic reproduction that reflects a system similar to human biological reproduction reflects the forces of the Law of Correspondence. In this model, galaxies aid our source in the process of creation. Just like Metatrons Cube. Essentially these systems also quintessentially align to universal law. The profundity of this expanded mental perspective lies in the fact that this macrocosmic pattern of galactic reproduction is occurring at all times in the cosmos even in this very moment. Somewhere in our microcosmic world, a child is also being born. The patterns of Nature have infinitely been repeating themselves since the dawn of creation. We can witness this awesome force of manifestation at work in the cosmos through a study of astronomy, astrophysics and cosmology. We can also now observe these systems through the immaculate optical capabilities of the James Webb Space Telescope. Images are now being produced and sent home for analysis by some of the most brilliant, highly educated minds of the modern scientific, academic community. This is a significant cause for tremendous excitement.

Has Halton Arp been vindicated postmortem? Indeed he has. The book of essays entitled *The Galileo of Palomar* is an excellent tribute. Additionally, with the recent paper written by 150 professional academics in all fields of scientific study, standard models of science are under the microscope. In a sense, Halton Arps vindication has already started to become a reality. Is the modern mainstream scientific community ready to submit his findings as evident proof of intrinsic redshift and galaxy-quasar interaction? No, not yet. While many scientists may agree with Arps conclusions, standard model science would have to bend and flex with great dexterity in order to permit experimentation and testing upon new alternative theories. Arp essentially proposed a process of galactic mirroring through self replication. Interestingly enough, his model reflects observable, orderly systems of macrocosmic creation in action at the highest perceivable levels. The concept also corresponds perfectly with the Universal Laws of Nature. In a world where nothing is truly a coincidence or a chance happening perhaps we can just call this fate.

Quasars emit light in the form of x-rays and microwaves, radio waves and highly charged electromagnetic plasma energy. Large host galaxies eject quasars, massive jet streams of light energy and galactic material outwards into the cosmic space around it. This scientific process correlates to the ancient primordial metaphysical system of creation responsible for generating the holographic universe that we live in. This primordial model of creation is in actuality, an immortal aspect of universal law that is always in effect and continuously creating life. Galaxies will never stop replicating themselves. They will infinitely continue to populate the cosmos with new stellar bodies all throughout the universe.

Why? Because as we already know, all things governed by the universal laws of our source are automatic self managing, self maintaining systems of creation. Humanity can witness this cosmic work of our creator and the spirit forces of Nature through very high powered infrared telescopes like JWST and Hubble. At this juncture in space exploration, modern science has access to the latest, most highly advanced technology that has ever existed in any human age or generation. NASA and all associated scientific observational teams now have more light to shine on the dusty mysteries of science than ever before. Independent scientists and professionals from various other fields of study also have access to much of this same new evidence as JWST images are publicly released into our academic world of research.

Creative forces of Nature are only assistants of the Absolute, our Creator. The galaxy, the quasar, the micro cluster galaxies and even the metaphysics of Metatrons Cube itself, do not create consciousness. The consiousness of our Source is the ultimate omnipresent force powering the alchemical engines that generate all scientific processes of Nature. This system is automatic, set in place, above the laws of man, incapable of human alteration and governed by Universal Law. By ejecting quasars out into space that evolve into cluster galaxies, gigantic host galaxies create stellar families. The cluster galaxies then evolve into larger galaxies and continue to reproduce new galaxies in the same manner as their parent or host. These stellar family units are scattered all across the cosmos many displaying high potential of unique compositional qualities of similar Nature. Other quasars stray and travel a great distance away from their parent galaxy during evolution. So are we seeing any quintessential correlations to Natural Law in the scientific systems and models proposed by Halton Arp? Most certainly. What we are exploring here, are the exact type of quintessential, spiritual correspondences that Halton Arp was also seeking to discover through his work. The propositions and theories set forth by the Galileo of Palomar and the evidence he has compiled, do align to known metaphysical systems of creation witnessed and experienced in our biological world. This information is of tremendous relevance in my own research of Arps findings as I am seeking quintessential connections within his body of work that help bridge the gap between science and spirituality. From that perspective, I am on the right track.

We have acknowledged the fact that Arps version of galaxy reproduction is a reflection of a similar process witnessed in human reproduction. The method by which we pro-create is in direct association with natural law and the scientific processes of Nature. Our biological family units created by our ability to produce offspring, reflect the endless luminous stars and radiant bodies that we behold in space. In other words our microcosmic process of reproduction correlates to that of massive galaxy formation. In embryotic cell division a single cell divides into a zygote, creating a vesica pisces. The divine spiritual light of the soul now has a doorway through which to stream forth into our biological world. A new life begins. Meanwhile in the Cosmos, another star is born. Everything is connected. The zygote represents duality as well as the emergence of light. A vesica pisces is created in this process of cell division. Just as it is in the construction of the dyad within the divine architecture of the seed of life. Through the vesica pisces comes the light of life in these initial stages of embryonic development. In the metaphysical origin model, this occurs on a macrocosmic level on the first day of genesis in the initial stages of primordial creation.

In Arps model, galaxies emit light and infant quasars from the heart of their divine center through massive energetic jet streams. All of these scientific processes stem from the consciousness of our Source. There is profound truth, limitless beauty and orderly intelligence embodied within these quintessential correlations. Universal Laws govern all different levels of manifestation occurring all throughout Nature at all times. All of it is an orderly, organized system designed, structured upon and generated through the omnipresent consciousness of our Source. The effects of natural laws in action, manifest what we think, see, choose or desire into existence. We energetically attract these conditions and experience the results as our reality. Promoters of the standard model paradigms will always defend accepted scientific core beliefs. Will modern science ever flex in mercy even slightly? Just enough to allow a bit of big telescope time and a little room for alternative, abstract theories? We shall have to wait and see. Consciousness elevation, spiritual human evolution and mental expansion, are born of bold moves such as this. Here's to hope that some day, mainstream science will step up to that plate.

As previously noted, God, our Creator and Source is known by many names. The Absolute, Allah, Yahweh, Jehovah, Amun, Atum, Ra, Brahma, Apollo, Zeus, Shamash, Divinity, the All and so on. Whatever you wish to call this divine omnipresence it is still all the same Almighty Source energy. The way that we view or address our Creator normally begins in childhood by way of cultural, social and family influence. After one reaches an age of reason, free will takes over decisions such as this quite naturally. Through growth and experience, our independent systems of belief are designed to eventually take precedence over the initial instructions and habits inherited from our family. Yet, it's imperative to note, that at the end of the day all positive moral, loving roads to God lead to the same God, the same Source, same creator, no matter what you choose to name it. This divine force doesn't really care what name you call it by. Our Source just wants us to remember to call. Imagine that. Consequently, in this light, all judgement we see in religion can instantly cease and desist, be rendered unnecessary and then disposed of. We needn't be overwhelmed by the choices of religion that we are offered as a race. The immaculate light and unconditional love of our Creator simply resides within every spiritual tradition that is pious, humble, ethical, just and morally good. God also resides within all things that are loving, descent and pure. Grains of truth have been planted in the soil of all major global religions and faiths. True devout Christianity, Eastern philosophical, esoteric, mystical, new age spiritual and even ancient tribal traditions were originally designed to resonate spirituality and love. Honest, humble, loving faiths, hope, prayer and dedication to our Source generates positive, elevated harmonics no matter which ethical or moral path you choose.

In the public academic domain there is a lot of good knowledge intertwined with a ton of bad, misleading and outright incorrect information. Researchers are put to the test of weeding through mountains of data utilizing conscious, rational, cognitive ability to discern right from wrong and the truth from a lie. We all have a built in moral compass that points to true North and recognizes truth when confronted with it. We all know the difference between bad and good. In regards to the information and knowledge that we acquire, we have been granted the free will right, to take what we want and leave the rest behind. This is how each of us builds our own spiritual, ethical, moral belief systems. As we go about doing this, we are obligated by Natural Law to remember proper human conduct, the 10 commandments, the golden rule and to abide by the laws of the Universe. All these codes of conduct are good, accurate and correct. These rules and guidelines are the heart, core and foundation underlying all major religions and faiths. These energies bind us all whether we are aware of this or not. Thus, it is imperative that we pay close attention to immortal, universal wisdoms such as these.

Look into the cosmos. Behold the light and you will see,

that creation did not end after the 7 days of Genesis. It is still happening. Witness the functions of the cosmos and the scientific processes of Nature in action, here on earth. Creation occurs at all times everywhere and this is plain to see. Existence is a fleeting blessing. We are spiritual beings having a human experience. Our souls are bound within biological earth vessels. Our physical vessels are comprised of 70% water. Trillions of molecules and cells are floating around in this highly charged medium in a constant state of vibration. All throughout Nature, creation continues every day. Every hour an egg hatches, a seed sprouts, a frog egg becomes a tad pole, a butterfly springs forth from a cocoon, a child is born, a new star bursts into existence and so on. The ball is always rolling. Creation never stops.

Here on Earth, humanity can be seen as a miniscule manifestation in comparison to all other things that exist in the universe. But we are by no means insignificant. Like quasars and galaxies, we too are creators. What shall we manifest as a race? What loving, harmonious energy can we as a species, raise to help make this world a better place to live in? It's up, to each of us to decide what our part in that mission actually is. But we always get to choose. That is what free will is about. We all have a moral compass to guide us for a reason. What we conclusively do know about the Absolute, our Source, from sound esoteric sources, knowledge of quantum physics and geometric archetypes such as Metatrons Cube or the Flower of Life, is that the same fractal patterns of creation repeat themselves all throughout the physical plane. These infinite processes, are uniquely bound by Universal Law and the spirit forces of Nature. Everything is connected. All is equal. All is One. Developing even a basic understanding of these truths, shows us that there is no logical reason for our tiny little planet to be the only earth like planet where biological life and humanity exist.

Humans are products of Nature produced through systems of creation in a constant state of infinite repetition. That being the case, it is safe, logical, rational, realistic, intelligent and very wise to assume that we are not alone. In fact, to assume such a thing would be a gross oversight and an outright rejection of sound reason. Are we even alone right here in the milky way? Is this all there is to the human race? Is there anybody out there? The odds are strongly stacked against the likelihood that we and our earthly abode are a singularity. In fact, NASA has already found planets out there with conditions similar to our earth and yellow dwarf stars, like our sun that are required to sustain biological life on such earth like planets. In regard to space exploration in our modern age, the scientific fun has only just begun. In metaphysics our solar system is in reality, a model solar system after which other solar systems are designed. This is so, as everything in Nature is a reflection and process of repetition. On a macrocosmic level we can look at our entire solar system as one small part of a massive, interconnected masterpiece of grand, divine, magnificent design. An intelligent hologram in an infinite state of vibration or motion. Through an understanding of the scientific processes of Nature and the primordial systems of creation bound by Natural Law it becomes obvious that our human race a microcosmic piece of a macrocosmic puzzle. As is our solar system. If one lone, yellow dwarf star can manifest so much life here on earth, imagine what all the rest of them are out there doing in the cosmos. Viewing our existence from this vantage point reveals the vastness of the great creation of our Source and how much more we truly have to discover and learn about our place within it as a species.

Everything we see in existence is part of one colossal interconnected system of divine omnipresent energy that is primordially composed of consciousness and light. We can expand our awareness and raise our consciousness levels to higher states of vibration so that we can internally perceive the whole of it from the cosmic egg right down to the egg of a frog. To perform this task we must open our hearts and minds to the light of truth. In this way we can truly come to understand and acknowledge the privilege and great responsibility that it is to be human. We are Spirit forms that are here on earth to experience biological existence in illusionary physical form as the master race of the animal kingdom. That being said, humans cause more destruction to the planet, to each other and to the innocent animal kingdom than any other creature or mammal in existence. Does anyone see anything dreadfully wrong with this???

We are divine manifestations composed of spiritual consciousness and the immaculate unconditional love of our Source in the flesh. We are actually here to reflect the characteristics of our Source and express the loving energy of our Creator outward through our heart chakra, into the space around us. It is within our heart that our connection to our Source resides. The vehicle used to arrive at a place of light, knowing and true wisdom is not a tour bus driven by our objective ego self. It's a mach jet piloted by our spiritual inner self. As long as ego runs the show, the wheels on the bus keep going round and round and getting absolutely no where. Ignorance is not bliss. Knowledge is true freedom. Our presence here on earth is both a privilege and a responsibility. This applies to every one of us. As such, we are all called to duty in regard to protecting and preserving our planet. It is everyone's job to cultivate, care for and protect Nature from the careless damage caused by our living habits and the less spiritually developed members of our race.

It is the duty of the strong to protect the weaker aspects of Nature. This includes fellow members of humanity and second density beings within our plant, ocean and animal kingdoms. Our biological vessels and physical capabilities were designed and equipped with all necessary faculties required to go about doing this privileged work. We are all participants in this quintessential operation from birth onward. We are to conduct this work of our Source as willing participants quite naturally, with unconditional love during our time here on earth as guests. Love is the core essence of our Source. This energy is automatically inherant within our human, spiritual composition. As living reflections of the macrocosm, we are all one star family. Just like all the cluster galaxies and quasars surrounding massive active galaxies. Our milky way galaxy alone contains endless, infinite numbers of stars, cosmic phenomenon and celestial bodies. According to the United Nations, our planet now contains over 8 billion people. All of us, are part of *One* interconnected consciousness.

 In regard to Macrocosmic processes occurring in our universe such as the true functions of quasars, galaxy formation and active galactic nuclei there is still so much essential study for the modern scientific community to conduct. So many stones have been left unturned and so many new ones are now unearthed on a regular basis since the 2022 launching of James Webb. Many significant observations have been made by professional scientists from various academic fields that require more research and experimentation. There remains so much to explore. Great minds have rendered great theories that have been left behind for humanity to ponder. We are free to acquire access to this public knowledge through study and research as we see fit. So many mysteries of the universe remain unsolved. It is humanity's job to resolve this issue.

 Astronomer Halton Arps entire body of work is filled with logic, rationality, painstaking research, reasonable conclusions and what appears to be sound intriguing evidence. In reality, all he did was flick off the tip of the scientific iceberg. Having researched some of his work, I see much intelligent clarity in his discoveries and observations.

Within his research, I found some of the quintessential connections to universal law that I was looking for. Arp proposed existing observable models of science that reflect microcosmic aspects of creation. The mission involved in the creation of this 3 book series *As above, So Below* is to discover correlations between spirituality and science. Halton Arp placed plenty of practical undeniable truths upon the modern scientific table for further analysis and review. The bases were loaded but standard model science did not step up to the plate and abandon the fundamentals of known accepted science. Instead his findings were classified as non-compliant with standard models of science. Halton Arp never made it into the scientific hall of fame for these amazing discoveries. Nor was he ever given big telescope time at Palomar or Wilson observatories to explore his discoveries further. Regretfully. His legend however lives on and his life's work is still part of our academic world. It is available to anyone who wishes to conduct further research and experimentation upon it. James Webb our golden eye in the sky will continue to lead humanity towards a new scientific frontier. Let us embrace the strange, the unusual and the abstract.

As previously noted, the origin ejection of radio lobes has long been accepted by modern science. X-ray jets are academically known to be found at their core. X-ray sources in the famous radio galaxy 3C295 are indeed indicated to be in a state of active ejection. Many members of the scientific community agree that further study and research of this magnificent, perplexing cosmic phenomenon, should be conducted. Profound aspects of Nature such as this are begging for our focused attention, further acknowledgement, academic interest and intelligent recognition. As previously mentioned, in metaphysics all things are here by way of orderly, intelligent design. Nothing is considered random, chance, coincidental or accidental. All is part of a grand, immaculate, self replicating, self maintaining template.

The truth is out there and in many cases it is often hidden in plain sight. It is impossible to deny the profundity an impeccable quality of the illuminated work of our Source. Our sights must be firmly placed upon seeking light if we truly wish to find it. Initially, we must expand our perceptive awareness and elevate our state of consciousness in order to even see it. At the end of the day one thing is certain. Demystifying science will facilitate profound, superlative comprehension of our purpose here on earth. Discovering the many spiritual scientific functions of our universe as they correspond to Universal Law is key. Unraveling fascinating mysteries such as these is our intellectual duty and inherant human birthright. By solving perplexing cosmic riddles we contribute to the spiritual evolution and consciousness expansion of our race. In this way, we help bridge the gap between science and spirituality.

CHAPTER 11

A Simple Overview Of Radionics And Electromagnetic Morphic Energy Fields

Radionics therapy was designed by a German research team at the Institute of Radionics Therapy in Germany (IRT). The technology used in radionics has the ability to locate a living organisms bio-frequency code. The operator then transmits the required signal needed to heal the chosen organism in distress, in the direction of their subject. The transmitted frequency is known in radionics as the *healing code.* Once the bio-frequency signal reaches the biological organisms resonant bio-field, the energetic light code is absorbed and the and the patients healing process begins. Utilizing this method of light code transmission through radionics, the IRT was able to heal entire forests and ecosystems as well. This information, is great news for mother earth. In addition, humanity now has access to the modern new Biogeometry science patented by Ibrahim Karim of Cairo Egypt which is comprised of ancient Egyptian energy work techniques capable of healing entire communities. Morphic bio field energy is very powerful in many places on earth where electromagnetic ley lines are present and even more highly charged on places on earth where ley lines meet and cross. These spots create highly energetic terrestrial and spiritual X points that often emit very powerful healing energies. These forces affect all biological life in the surrounding areas.

Magnificent temples of great magnitude and powerful, quintessential structures have been built upon these divine sacred energy sites by members of the human race within every generation for millennia. As we all know, some of these primordial antediluvian ruins are still here today. The megalithic ruins are of the most ancient sacred temples and divine constructions of this great cycle of sidereal celestial time. They were built by ancient ancestors that came long before our current age and even before Noah's flood. These temples that were built by our ancient ancestors were designed in such as way as to capture and retain the energetic healing forces and spiritual energy of Nature that emanates from these specific locations. These ancient sites do indeed have expansive, powerful divine energy. It's a commonly known fact that sacred ruin sites have a tendency to promote and facilitate spiritual awakenings, transformational experiences, telepathic encounters with spirit, elevated consciousness and so on. For those of us sensitive enough to receive such signals these sites are well equipped to infinitely transmit their frequencies. That is because the scientific processes that are the force behind holy temple construction. Ancient temples were built to embody the power of the repetitive infinite forces of creation that are governed by Universal Law. All living biological organisms can sense the spiritual energy and unseen mystical forces retained at these sacred sites and inside such holy temples.

The energy that is present in these areas is a sympathetic frequency of higher harmonic resonance. This energy promotes expanded awareness, higher states of consciousness and states of mental transcendence. The many spiritual experiences that commonly occur at such places such as the Giza pyramid or Stonehenge are far too numerous to count. In the science of Radionics a healing code is transmitted from a controller to a living biological part of Nature. The signal can be transmitted over vast distances of space and time. Codes of this nature have been used to heal biological lifeforms such as forests and eco systems. This is a very reliable scientific method of treating any living organism that is 1000's of miles away from the location from which the code was transmitted. This amazing revolutionary science verifies the critical significant role that frequency plays in our existence in our material world. The radionics technique is a proven scientific method of tapping into light forces contained within a subtle energy code. The code is then transmitted outwards in the direction of the biological part of nature that is to receive the transmitted frequency code. The energy of both sound and light move through time and space in the form of a wave which can be measured using the Doppler effect model. Electromagnetic energy moves in the form of an electromagnetic transverse wave or as a particle. Sympathetic resonance is at the heart of this amazing, phenomenal form of scientific energetic information delivery. The metaphysical symbols used in Radionics are of primordial construct. Encoded within them is immortal information that relates to the many universal scientific processes of nature and Creation. Ancient energy symbols also known as light codes resonate with all levels of existence from macrocosmic to microcosmic levels of manifested existence. Fundamentally they embody the spiritual forces of Nature.

Healing codes or light codes come in many different shapes, sizes and forms. They are primarily composed of light and frequency in various states, patterns and rates or speeds of vibration. Metaphysical healing codes are found to be encrypted within the fabric of the scripts of ancient texts, rune stones and geometric symbols. These codes are commonly known Alchemical archetypes such as cymatic energy forms, mandalas and even crop circle formations. The Flower of Life and the many archetypal components of which it is composed contains the most powerful healing codes of creation. This symbol is known globally to all generations of the human race. The Flower of Life is the foundation for Metatrons Cube and as such the underlying structure of the Isotropic Vector Matrix. All light code signatures required to build these divine quintessential aspects of creation are intricately encrypted within the 3D Flower of Life. It is the foundational building block of our holographic universe as it exists within the Cosmic Egg. The 3D Flower, the IVM is the most powerful force for generating life within creation on the physical and cosmic planes of existence. As anyone can learn from studying the energy line mapping work of Becker-Hagans, the current icosa-dodecahedron crystal grid is evolving into the Polyhedra 120, which is a 120 point star. It's architecture is composed of icosa-dodecahedral geometry with 12 pentagonal faces. Each of the 12 faces bears a 10 pointed star. Thus, 120 points completes the whole. This is the force underlying the great global awakening.

 The architecture of this divine sacred geometric form contributes to spiritual awakenings, higher levels of consciousness, unity, oneness, spiritual evolution, expansive energy, compassion and universal love. This ascensional energy can already be felt, sensed as well as absorbed by all living biological organisms and human beings on the physical plane. This omnipresent, quintessential, scientific phenomenon of enlightenment is the force that is contributing to what is commonly becoming known in New Age traditions and metaphysics, as the Global Ascension. Many members of humanity are open to these higher vibrational frequencies. This effect is clearly evident by the growing interest in the New Age community and other spiritual belief systems. The Isotropic Vector Matrix has 144 triangula faces . Each triangle is 180 degrees. Thus when we multiply 180x144 we get the amount of years in a great processional cycle of 25920. The IVM is the 144 pointed geometric star that generates our holographic universe and all the things we see in existence. in this way, 144 is a very sacred number. The IVM is a God Star. It's components consist of the most recognizable archetypal universal symbols in all of creation all nested within one another. These archetypes are popularly known as the platonic solids.

One of the most relevant symbols encoded within the IVM is the Pentad. This is a 5 sided geometric form. The 5 pointed pentagram star forms upon the face of the Pentad. This 5 pointed star is the most widely judged and misunderstood of all ancient symbols used on Earth today. Of all spiritual traditions combined, this symbol is abundantly popular in human culture and worn by many. In today's modern society, it has become nearly equal in popularity to the symbol of the cross. In Nature there is no evil meaning, intent or connotation attached to this symbol at all whatsoever. The minds of man have however added a satanic element to this divine symbol of creation which quite simply was not the intention of the Absolute, our Creator and that is a fact. Period. The entity known as Satan is of religious significance and does not exist in metaphysics as it does in the religious doctrines of churches on earth. Satan is a construct of man, not Nature and certainly not of science. Nope. No devil in science either. Evil forces such as this exist within the minds and resonant energy fields of humans who create, empower and project evil thought forms. Does this mean that you shouldn't be a good person? That you can just go right ahead and sin all you want to and defy the laws of nature? Nope. It means that you are even more bound to proper moral conduct and good ethical behavior because you know the difference between right and wrong behavior. Primordial constructs of Nature resonate the unconditional love of our Creator, not the opposite. These forces are generated through the Laws of the Universe and governed by the same spirit forces of consciousness. The Pentad is formed in the 4th oscillation of the Seed of Life. The star is born when the tetrad, the 4 principle aspects of biological life is united with the force of spirit. The tetrad embodies the forces of earth, wind, water and fire combined with the force of quintessence or spirit. When using the solar spectrum as a guide, this oscillation correlates to our yellow solar plexus energy center and the sun. The most globally recognized symbol for the sun is the divine sphere with a dot in the center. The exact same symbol that represents our Source. The symbol that corresponds to the omnipresent essence of the Absolute. The geometry of the pentad is one of the forces that makes human beings literal stars. The famous image of Leonardo Davinci model of man with outstretched arms and spread legs within a circle and a square is critically significant symbolism in our physical world. The cube in the image represents the four principle platonic solids that compose all biological matter including our earth and earth vessels, our bodies. Also the 4 lower points of the star. The sphere represents the force consciousness of our Source. Quintessence, spirit soul consciousness and life force, are encompassed within the 5th point of the pentagram star. The 5th element quintessence corresponds to the top, upward aiming geometric point of the Pentagram star. By way of intelligent design, this energy points upward towards our Source respectively. The top point off the star represents our head and crown chakra.

Together these forces represent all 5 geometric platonic solids. In addition, these are the 5 geometric archetypal forces that comprise our earth vessels. In other words, our physical bodies. During each incarnation our soul and consciousness are confined to a biological vehicle in which our soul must reside and travel while here on earth. In this way each of us is a star. However, there is also a Merkaba star energy within our human composition that we can access through our heart chakra. This divine energy resonates with our chakra column and activates our Merkaba light body. So ultimately, as human biological lifeforms, what we truly are, is a star within a star. Each and every one of us is an aspect of the holy divine energy of our Source and as such, a miracle of Nature. Our biological earth vessels are the essential requirement necessary to house our soul spirit while we are incarnated on the physical plane as physical beings having one of many temporary human experiences. Each point of the Pentagram represents not only the 4 principle biological energies found all throughout creation on the physical plane, it also represents the cardinal points of North, East, South and West. The 5th point respectively correlates to that which is above us in the realm of spirit. Finally, the 4 points of the pentagram represent our arms and legs and the 5th point up representing quintessence is at the position of our head where our consciousness functions generating our brainwaves. These 4 principle aspects combined with the 5th force of quintessence composes the essential fabric of human beings. The force of our Creator, the sphere in Leonardo's model of man as a star, surrounds the symbol of man enclosing him within the safety of the boundaries of our Source. Essentially when the circle is placed around the Pentagram this symbology represents an element of our Source, divine spirit and ultimate protection.

As the Seed of Life expands into the Flower, in 60° outward oscillations the 5 platonic solid archetypes of creation are generated. These 5 vital energy forms, with the addition of the hexad are the essential energetic ingredients required for manifestation in the process of crystallizing all light energy into biological matter. As is commonly taught in metaphysics the pentagram star has 5 points each measuring 72°. In ancient traditions, 72 is a sacred number associated with the name of God. The Golden Ratio, PHI 1:1.618 is born within the construction of the Pentagrams 72 ° angles. The sacred fractal PHI can be found, witnessed and observed within the constitution of various lifeforms right here on earth. These correspondences are reflected everywhere all throughout biological creation and Nature as we perceive it. The holy number 72 is also associated with calculations of zodiac ages and tracking the sidereal zodiac processional cycles. The zodiac ring of stars forms a 360° wheel in the Cosmos. Each zodiac sign is said to take 2160 years to transit into the next sign. This ring of stars is of course, known as the celestial sidereal zodiac. The great ring of fire residing in the lofty skies. It is said that each 2160 year zodiac age occupies a space of 30° on the zodiac wheel. It takes 72 years to transit 1° upon the wheel. One complete 360° revolution around the zodiac wheel of stars would as such take 25,920 years to complete. This celestial cycle is also commonly taught in science as the cycle of earths processional wobble.

The processional wobble of our earth is said to be the cause of the 23.5° angle tilt of our planet. Modern science has long since revealed the axial tilt of the earth to be slightly off as opposed to sitting straight upright in space. This causes a slight angle of the equatorial plane and equator by 23.5°. The 25, 920 year cycle of the wobble was also meticulously recorded in ancient Mayan long count calendrical systems. It is taught in modern astrology that the zodiac band of stars naturally transits 1 ° every 72 years. A 360° rotation of all 12 zodiacs at 2160 years each occupies a space of 30° on the 360° zodiac wheel. 30° takes exactly 2160 years based on 1° of stellar transit every 72 years. 72×30°=2160 years and 2160×12 zodiacs=25920. Regardless of the accuracy of the equation, all ancient philosophers knew that all 12 zodiacs were not of exact equal proportion or size. Some were larger while others were smaller. This critical factor would alter the timeframe in which a zodiac travels though our locality of space. However, despite this significant stipulation, in today's modern age, the processional zodiac is calculated as explained above . So be it. This tool is the only known model that modern astrology offer as a means to calculate zodiac ages. Unfortunately, it may not be very accurate.

In recent years modern sidereal astrology teachings

confirmed the existence of a 13th Zodiac residing at the center of the zodiac wheel. The zodiac was given the name Ophiuchus. However, this constellation was very common knowledge in ancient Babylon and even further back in time. Ancients knew this highly electrically charged binary star system in the constellation of Draco as Alpha Draconis. This constellation generates the electric torus of the polar configuration. The name Thuban has also been applied to this constellation . Thuban was a component of pole star configuration or Central Sun during the last golden age of man over 3000 years ago in the time of Atlantis. The Central Sun was commonly known as Helios. This time period goes all the way back to the epoch before Noah's flood. It was also known of by other ancient civilizations as Kronos. As a result, today Alpha Draconis is still associated with the constellation Taurus. The previous golden age of man dates back to the pole star Daneb of the constellation Cygnus during the age of Leo. Prior to this, the pole star was Vega of the constellation Lyra during the age of Scorpio. Today, during the age of Aquarius the known pole star is of course, Polaris. All ancient people had a deep understanding and highly educated knowledge of these celestial cycles.

Further information about this powerful phenomenon can be obtained through a study of David Talbotts work on the reconstruction of the polar configuration known to all ancient ages as the Central Sun. The polar configuration at the center of the zodiac wheel produced a visible second sun in the ancient sky. A comprehension of Talbotts research can help anyone develop a comprehensive understanding of this ancient knowledge. His massive body of meticulously complied research can be found at www.thunderbolts.info. His following highly enlightening videos can also be found on YouTube *Remembering the End of the World,* which describes cataclysms of past, *Mythology as Undiscovered History* and *Symbols on an Alien Sky* as well as *Thunderbolts of the Gods.*

There is of course so much more scientific information put forth to study and process once one enters the academic mind of David Talbott. For amazing revelations into past pole star history and the ancient civilizations who openly knew of this knowledge read best selling author Freddy Silva's book *The Messing Lands –Uncovering Earths Pre-Flood Civilization.*

When all 5 of the 72° points of the Pentagram are multiplied together the sum total is 360°. The circumference of a sphere. This is the shape and divine measurement of the original quintessential symbol, the sphere of the Absolute. The Pentagram enclosed within the sacred circle completes the symbol as a divine representation of creation. Consequently, bearing this in mind it is no surprise that this symbol became the archetypal light code symbolism of the living, breathing, human biological reflection of our Source. In the scientific processes of Nature all manifested energy forms embody the sacred structure and force of the original primordial sphere. In addition everything in nature is structured upon endless returns of the octave of light as energy moves throughout the spectrum at differing rates of speed and in various states of vibration. The law of octaves teaches us that all energy oscillates outward expanding in the form of a circle that infinitely returns to the divine center to begin again. This process is an endless fractal pattern of creation formed by repetitive 7 step spirals of energy that constantly, naturally return to their source in an 8th step upon completion, only to reinitiate. This is the geometric force of infinity at work in Nature.

After multiple incarnate experiences our Souls are said to complete their journey of growth and spiritual evolution. Only then do our souls return home to the realm of the Absolute, the abode of our Creator. All things in our physical plane are formed of light and electromagnetism in multiple states of vibrational frequency. These forces manifest all energy into existence through the functions of Metatrons Cube. Through the observant academic eyes of science we can truly behold the magnificent work of our Creator. Nature is beyond spectacular. Everything in existence is fundamentally composed of light, consciousness and energy in motion. The highly advanced methods and techniques used in Radionics are uniquely successful. This is because everything in biological existence is comprised of a complex combination of the same energy that is utilized in the healing code symbols. Every form of living biological matter emits light, frequency and electromagnetism through a Morphic energy field that encompasses and surrounds it. Some people are sensitive enough to energy that they are capable of observing auras of light and emissions that exist around certain biological lifeforms. Our aura is an energetic field of chakra light emitted by our sympathetic astral body.

The colors of our aura and chakras are composed of invisible wavelengths of electromagnetic energy which is always flowing outward from our biological bodies, our sympathetic energy centers, brainwaves, electromagnetic torus and so on. Regardless of whether we can see it or not our energy field is always there and always emitting energy in the form of light. Those who have the ability to see or feel this light can offer fundamental insights into a persons physical and emotional and spiritual states of being. Every human, plant, animal and so on are surrounded by a Morphic energy field. This energy field is a primordial construct and model that is reflected everywhere all throughout creation. The shape that these fields form is that of a set of concentric rings comprised of alternating magnetic and electric energy. All over the earth ancient human civilizations have built endless copies of concentric ring structures and temples to honor and reflect this sacred scientific energy form. The quintessential functions and structure of this electromagnetic energy are omnipresent models of manifestation.

Even the mystical capitol city of the lost continent of Atlantis, Poseidonia was known to be laid out in the form of concentric rings. Discovering this ancient ruin site has been the relentless pursuit of many explorers to date. In fact, concentric rings symbolism can literally be found all over the earth. It has been discovered on various ancient rock carvings and prehistoric petroglyphs. It can be seen in native art found on cave walls and areas of the grand canyon. The globally known concentric ring symbolism is commonly known of in various professional fields of academic study. Geologists, Archeologists, paleontologists, researchers, explorers and so on, have in fact uncovered various natural ring like structures as well as man made concentric ring ruins such as Arkaim in Russia and Stonehenge in England. Even crater impacts can leave behind concentric ring like destructive formations in their wake. The Richat Structure once mistaken for the ruins of Atlantis was ultimately forged by nature. The concentric ring ruins known as the Moray Ruins of the Inca in Peru also replicate this immortal energy form. This symbolism does not describe a random construct of creation or occasional occurrence in Nature by any means. Multiple functions of Nature automatically create ring like formations as a natural effect. Concentric ring like formations are universal symbolism known to every generation of humanity. Morphic resonant energy fields and concentric ring design is also seen in Nature in the form of symbolism associated with earthquakes, seismic activity and geo-electromagnetic and magnetic pole shifts. Planets within our solar system orbit the Sun in perfect concentric rings. This intelligently structured model reflects the same sacred, primordial symbolism within its divine construct. Randall Carlson has produced a 16 hour documentary uncovering the potential location of Atlantis on the mid-Atlantic ridge on his You Tube channel *Kosmographia*. This information is highly educational for anyone seeking to learn more about what geologists and oceanographer already know about the ruins found on the North Atlantic ridge.

Metaphysical scientific processes of Nature literally create the concentric ring energy form all throughout creation. Drop a pebble in a pond. The resultant effect of this cause will always be the formation of evenly spaced concentric ring like ripples on the surface of the water. We have all seen this effect. The rings will always naturally form around the point where the stone fell, emanating outwards from the divine center point until the rippled wave loses its energetic force of inertia. At this point the surface of the water returns to an undisturbed state. Ring like designs can be found everywhere in nature that energy exists. Electromagnetic lay lines run all throughout the earth and are comprised of two energy pathways of electric and magnetic forces. In many locations on the earth these energy lines merge to form significant cross points creating areas of highly charged spiritual energy which ancients have utilized for healing purposes, to harness earth energy, for ceremonial rites and divine worship for millennia. This has been a well known fact for centuries. Whatever structure is erected on sites such as this will emit a very powerful resonant bio-field energy.

Modern maps of energetic ley lines and crystalline geometric grid lines such as the Hartman Map, R. Buckminster Fuller Dymaxion and the Becker-Hagans Earth Crystal Grid Map can be found at on the following website: www.earthacupuncture.info/earth_grid.htm. The Morphic energy fields in specific areas on earth were considered sacred by ancient civilizations and of great spiritual significance. Great ancient temples captured the healing forces, powerful resonance and sympathetic vibratory ascensional energies of the earth. These temples were built on these highly energized sacred sites by ancient builders who had full awareness of the metaphysical energies housed there. In Russia, Hartman and Curry energy lines were referred to as the *Matrix of Cosmic*

Energy. This energy, along with its symbolism are complex primordial constructs of Nature found all across the surface of the earth. The electromagnetic torus of biological organisms forms a central column that moves through the life form. This energy field generates a ring like or doughnut shape of electromagnetic toroidal energy. This energy encompasses the biological organism in an egg like barrier. This energy field extends and emanates outwards from the biological life form in the familiar concentric ring like pattern. It can extend up to 6 feet outwards from the central point which is largely dependent on the height of the biological organism.

Morphic energy fields which surround each biological manifestation come into contact with the resonant bio-field emanating from other biological organisms in the environment around it. When we are physically within close proximity of another person our Morphic bio-field comes in contact with their resonant bio-field. When this occurs, the two sets of concentric energy rings that comprise the bio-field overlap. This over lapping of energy fields allows for energy and information transfer. Consequently, when we are in the resonant energy field of another person we can instantly feel this outside energy. We pick up energetic signals off other people and animals as well. These signals can often tell us immediately whether we could have good or bad rapport with that individual or if an animal is friendly or dangerous. We can feel instantaneous abrasive energy or instant kinship. Energies such as these arise quite naturally whenever our resonant bio-field comes in contact and communicates with any other resonant bio-field of any other living thing found in Nature. Now that we know the spiritual significance of it, go ahead, hug a tree. It certainly won't hurt anything. Bio-field energies in the form of concentric ring like structures can be found everywhere in our biological world of matter. They are spread all throughout all kingdoms of material things.

The energetic structure and composition of the concentric rings can be seen as alternating red and blue energy lines. Electromagnetism is comprised of these two forces of light energy in motion, undulating in an alternating pattern of perfect rings. The red rings are comprised of red XY light which is fundamentally electric energy. The blue rings are comprised of blue XX light which is primordial magnetic energy. Electromagnetism is comprised of these two forces in combination. As these Morphic bio-fields emanate from biological aspects of Nature, these dual forces automatically alternate in perfect order outwards and away from the organism emitting the bio-field. As noted, our bio-field energy can extend outwards up to 6 feet or slightly further for a taller person who is over 6 feet high. At the divine center of these sets of concentric rings is always where we find the biological lifeform which can be symbolized as the divine dot at the center of the sacred sphere. In this way we see yet another corresponding reflection of the macrocosm within our microcosmic world. Sound familiar? It absolutely should. We are made in the image of our Source. Indeed we are a reflection of the Absolute. The primordial sphere with a dot in the center residing at the heart of the Seed of Life. In regard to quintessential models of creation this concentric ring model resonates with the Natural Law of Correspondence with precise accuracy. Acknowledging this information is key to developing an understanding of how all macrocosmic and microcosmic aspects of creation are intricately connected. Natural Law governs all scientific processes of creation conducted by the spirit forces of Nature at all times.

From the dawn of time and the Genesis onward every manifested thing in creation is a biological copy or reflective quality of another original biological model found somewhere else in creation. This is the fundamental building block of Natural Law and the foundation stone of the Law of Correspondence. In metaphysics the primordial model of the Cosmic Egg and the geometric components and functions of the Flower of Life operate symbiotically in perfect combination and unison. These forces generate the holographic universe we live in. As a model of creation do we see the fractal patterns of the geometric energy of the platonic solids reflecting within various observable manifestations in nature? Yes we most certainly do. What we do not see happening is the explosive random energy of creation events such as the Big Bang recurring all throughout Nature. On a spiritual scientific level, this technicality poses a problem. From a metaphysical perspective, a working primordial model of creation must accord with Universal Law and intelligent design. In metaphysics the primordial model of creation must initially meet the requirements of the Law of Correspondence in order to be functional all throughout creation. We live in a universe that is built, manifested and woven together through geometric, fractal systems of infinite repetitive cycles and patterns of creation. Under conditions such as this each living manifestation becomes a reflection of the next one. These scientific processes of Nature explain the functions of the Law of Correspondence, the Law of Rhythm, the Law of Polarity or Duality, the Law of Vibration, the Law of Cause and Effect and the Law of Mentalism. In our biological world we can find and observe many living biological life forms and energy structures whose constitutions reflect the qualities of the golden ratio, the Fibonacci sequence, platonic solids, sacred geometry and harmonic frequencies. The Big Bang origin model simply does not accord with the Law of Correspondence or any other metaphysical Universal Law as we properly know them. This inconsistency poses a logical threat to the standard origin model.

 Understanding resonant bio-fields of concentric rings that form around biological matter is the essential knowledge required to comprehend how we receive and send out electromagnetic energy. We are all familiar with the physical and emotional sensations known as good or bad vibes that we receive from our environment and other people. As you now know, the energy we feel originates from the resonant bio-field that surrounds them as it comes into contact with our own bio-field energy. All Electromagnetic resonant energy fields will always transmit energy between them when they come into contact. We can use this force to sense and channel information found within the energy fields of other living things. These Morphic fields exist everywhere in Nature that matter can be found. Concentric rings of electromagnetism form around people, plants, animals, flowers trees,planets, insects, crystals, stones, mountains and so on. We can feel this foreign energy even if we are blind, if our eyes are open or closed and even when we are navigating our way around in a pitch dark room.

We are always picking up this Morphic energy off of the things around us. This occurs any time anything else occupies the space that we are in. In other words, at all times. As we have already learned, the Morphic bio-field energy takes on the concentric ring structural formation beginning with the biological organism taking the position as the dot at the perfect center of the initial ring or sphere. In regard to the laws of Nature, this is of course a perfect reflection of the Absolute, our Source. We interact with these resonant bio-fields that form around everything in existence. Even those of inanimate manifestations like stones or mountains. In fact many mountains on earth are deemed sacred mountains as a result of the energy lines and bio-field energy that exists in the mountains location. Thus, the ultimate essence and only known face of God, the divine sphere with its sacred unknowable center is reflected everywhere all throughout the physical plane and biological world of matter. To acknowledge this is to comprehend how the powerful divine omnipresence of our source is found everywhere in creation. This model of manifestation perfectly aligns to the Universal Laws of Nature. Our material world is filled with reflections of our Creator. We have all been taught in spiritual and religious traditions that we are made in the image of our Source. Resonant Morphic energy field concentric ring like structures as they exist in nature, is only one of many examples in existence that continues to support this truth.

A vital detail to note at this critical juncture, is that this type of bio-field energy can only exist in an electric universe. This fundamental detail is an obvious, logical fact. Bearing this in mind, humanity can come to a much clearer comprehension of the powerful significant role that electromagnetism and vibrational frequencies play in our existence. Knowledge such as this also facilitates an intelligent awareness of how our resonant Morphic bio-fields can be affected by radionics signals. Utilizing radionics, energy healing codes can travel vast distances to treat a biological organism. This is a successful method of treatment despite the distant location from which the signal was transmitted. This is because everything in existence is composed of energy which in the holographic universe knows no boundary aside from the shell of the Cosmic Egg. Furthermore, at the core of all this energy, we find the force of electricity.

The healing codes administered in Radionics are universal constructs of creation. They are comprised of infinite energetic light codes that form the foundation of our universe and our existence as we know it. As previously noted, these sacred codes have been known of and used by healers and educators for millennia. These healing codes have been and always will be encrypted and hard wired within the energetic composition of sacred geometry, mandalas, Fractal equation patterns, cymatic frequency images, ancient alphabets, rune stones, the Architecture of the Flower of Life, Metatrons Cube, the platonic solids and the Isotropic Vector Matrix. They are found within the construct of crop circle foundations which is a key location to promote healing of the earth and humanity. The functions and processes of these quintessential forces comprise the light code energy symbolism used to create Crop Circle and Radionics cards. This is the foundational healing energy transmitted by the technology of Radionics. In a universe filled with biological life forms that are surrounded by electromagnetic energy fields, it's an excellent form of treatment. These concentric rings of alternating magnetic and electric energy are highly conductive receivers and transmitters of energy. This knowledge serves to further validate the immaculate interconnected oneness and unity that energetically binds all things in existence together on the physical plane. It also supports the undeniable truth and reality of the known scientific presence of electricity and its many functions. This powerful force is always in operation in our universe and in Nature all around us, at all times.

Finally, it's important to recognize the connections between Morphic bio- field ring like structure and the architecture of the Flower of Life. The sacred sphere is at the heart of both energy forms. A comprehension of the functions of these scientific systems lends tremendous corresponding credibility to the Cosmic Egg, holographic universe creation model. The electromagnetic energy that is dispersed all throughout our universe is primordially and fundamentally composed of light in a state of high vibrational frequency. The functions of electromagnetism automatically align to universal laws such as the Law of Vibration, the Law of Duality, theLaw of Polarity and so on. This heightened observational view, facilitates an understanding of how all things in existence are comprised of light transduced to frequency in a state of vibration or motion. This energy is further reduced to matter through the crystallization process via the functions of Metatrons Cube. Our consciousness and brainwaves operate by moving through a specific hertz range of varying rates of harmonic frequency. This automatic pre-set system of sympathetic frequencies activates our consciousness levels placing our brainwave states within the Epsilon frequency range of 0.05HZ which is below Delta to as high as 200HZ which correlates to Lambda brainwaves.

Our brainwave ranges each have specific qualities that generate specific states of awareness. Each brainwave frequency will activate different levels of consciousness, low or high vibrational states of mind, body and spirit and even healthy or negative mindsets that contribute to growth or illness. Learning how to gain control over our brainwave states utilizing intention, will and mind power is the most advanced energy tool we could ever wish to have access to. In the final chapter of this book we will look at a simple method that can be deployed to control our brainwave states and heart rate at will. Our brainwaves are comprised of a pre-set range of sympathetic frequencies that generate 12 known brainwave states. All of these brainwave states are active in our mind and body before we are even born. Our brain is essentially electric. Furthermore, our brainwaves are comprised of vibrational light in the form of frequency intertwined with the force of consciousness.

The frequency range of our brainwaves is a universal model that applies to all biological lifeforms equipped with cognitive rational thought capabilities. In other words, every member of the human race. Each of our human brain wave states corresponds to a specific set of frequencies within range of .05Hz-200Hz. Our chakra energy centers are likewise comprised of electromagnetic colored light energy which is fundamentally, frequency in a state of vibration. The varying rates of speed at which everything in creation vibrates creates the many different material manifestations that exist in our biological world of matter. Energy in a state of vibration composes all aspects of Nature that we experience as our reality. Frequency and light are at the heart of creation in our holographic, electric universe. This significant foundational knowledge is key to decoding the mysteries of our existence. For those seeking light, it is imperative to develop a basic understanding of the Morphic bio-field energies that surround all physical manifestations. This knowledge facilitates a comprehension of how bio-field technology combined with the practice of meditation can successfully lead to expanded, transcendent states of mental awareness and spiritual transformation.

CHAPTER 12

Reviwing The Benefits Of The Most Advanced Bio-Field Technology Meditation System Available On Todays Global Market

IAWAKE Technology was created in Utah, U.S.A by CEO of IAWAKE Technologies, John Dupuy MA and Pam Parsons Dupuy MA/LMFT who holds a degree in transpersonal psychology from J.F.K University. John and Pam are also the founders of the IAWAKE global community. John Dupuy holds a masters degree in transpersonal psychology and has been working with brainwave technology since 2004. He holds a bachelor of arts from Texas State University. Pam Parsons Dupuy has also been personally and professional working with brainwave entrainment technology for many years. To begin, let me lead you right to the source. The IAWAKE bio-field technology key program is called Profound Meditation 3.0 Full Spectrum. This amazing bio-field technology system can be found on the IAWAKE official website www.iawketechnologies.com . This program came into my life over 2 years ago. I made a commitment to the healing process and proceeded to actively use it on regular basis for a full year. This product review is a true testimonial of what this bio-field technology did for me. I believe it deserves a whole chapter in this book because of the uniqueness of this brainwave entrainment system. This program is a full spectrum brainwave bio-field meditation. This particular system stands out on its own from all other binaural entrainment systems available on todays market. This is a result of its original, unique, intelligent design combined with its very powerful impact.

The Profound Meditation 3.0 Full Spectrum biotechnology program was designed using digitally captured frequencies, computer software from numerous states of consciousness amplified 1000 times using patented IAWAKE processes. After using various binaural systems I can personally tell you that this programming is something entirely different from the norm. In regards to energy healing tools that offer us the ability to expand our consciousness, this product definitely delivers. The energy fields used to create this biotechnology programming have been thoroughly tested and proven to be healthy and beneficial to biological organisms using advanced Radiesthesia techniques. IAWAKE Full Spectrum Profound Meditation 3.0 uses energetic signatures that are proven to impart profound spiritual, mental and neurological benefits. Like all other binaural entrainment systems, this program is to be used with head phones. The bio-field technology embedded within the PMP 3.0 full spectrum system is so powerful that it causes a bio-field to form around the device that the audio program is being played on. The bio-field energy then extends into the space around you in addition to moving through you and within you. In this way the bio-field energy raised aligns and attaches itself to your particular bio-field signatures. The high vibrational energy is then able to remain with you for hours after the audio meditation session is over.

Personally, I can tell you that this program works. Using it on a regular basis has given me the ability to develop a powerful connection to inner self. I now have the capacity to draw forth wisdom and knowledge from this divine aspect of my body-mind-spirit complex. Developing a relationship with our inner self is essentially the greatest human mission during each incarnation. In absence of this connection we are nothing but physical beings having a physical experience. However, there is much more to existence than that. Overcoming lower aspects of our ego, outer objective selves, weaknesses, fears, anxieties, addictions, bad patterns of behavior and so on can only ultimately be accomplished when one is able to connect to inner self on some level. This is where our personal power to overcome our material ego originates. This is where our compassion, humility and capacity for forgiveness resides. Calibration of mental, emotional and physical faculties is necessary before one can achieve this. This programming initiates and facilitates a mental house cleaning and an emotional unburdening. During this process there is a direct confrontation with all demons that the outer self collected along the journey. All of these dark aspect of psyche must be cleansed out to make room for light.

All negative energy must be released. A clean white slate is critically required if you wish to fully achieve the immaculate results that can be attained through utilizing this biotechnology. We ultimately have the ability to achieve a state of consciousness and awareness that contributes to self mastery over mental and emotional faculties. We have the ability to perform phenomenal tasks such as transcending our consciousness to connect with the one unified collective mind within the vector field. The divine plane from whence our consciousness emerges contains all information in existence. While many systems on the market offer us this gateway to light, only one system that I have ever used actually delivered this. Every other binaural program that I used on a daily basis prepared my body mind and Spirit for the IAWAKE journey by calibrating all my energy centers to proper solar spectrum and brainwave frequencies. That initial step prepared my energy centers for easier activation allowing Profound Meditation 3.0 Full Spectrum biotechnology to work its magic. That is exactly what it did.

In general, our ego self is not to be entertained, amused or catered to. Additionally our ego often needs to be reminded that its not the boss. Whenever it is offered a crown it will put it on, wear it quite proudly and rule every aspect of our life quite willingly. But decisions such as this bury your inner self under a landslide of physiological debris so high that you can't even feel the spiritual aspect of yourself. For this reason there are hundreds of thousands of people that are completely unaware that this fundamental part of themselves, their inner self, even exists at all. Knowing oneself inside and out is the difference between being asleep or awake. If your desire its to empower, unite with and commune with this angelic essence within, then do not fight the releasing energy that is built into this programming. Removing and cleansing out this unwanted energy is a vital part of what Profound PMP 3.0 Full Spectrum biotechnology was designed to do. The releasing of this unnecessary energy can be overwhelming for those who have an abundance negative experience built up inside them from an entire lifetime of pain and lack of acknowledgment. Yet, it is a critical requirement that must be met in order to let this program give you the ability to unite with inner self and align to higher levels of consciousness.

Those of us who do not discover the presence of the spiritual light within and make that connection, will only have to experience more and more incarnations. The mission remains unaccomplished until the objective self learns to direct its energy inwards to find this powerful presence. Our sacred inner self resides within our heart chakra energy center. This includes our heart chakra, our Schumann causal chakra and our infinity causal chakra. These chakras can be seen in the charts in chapter 2, produced through research and experimentation, conducted by Dameon Michael Keller. These charts offer an accurate guide to follow based on correlating chakra frequencies to the solar spectrum. The question is, can

ProfoundMeditation 3.0 successfully embody both features by way of intelligent design? Can it initiate the release and bring the light to fill our vessels afterward? Can it aide us in transcending consciousness and connecting to the collective mind? Yes, yes and absolutely. This has been my expansive personal experience.

 Various frequencies are available on the market that have been designed using multiple different modalities of energy healing methods. That being said, none of them embody the bio-technology that is found within the architecture of PMP 3.0. It is imperative to inform you that prior to beginning my experience with IAWAKE technology I used various different frequencies and binaural programs for 1 full year. These were accurate solar spectrum generated energy center binaurals that properly activated the corresponding brainwaves for each specific chakra. I used these on a daily basis for a full year. The next step in my energy healing program was to incorporate Profound Meditation 3.0 Full Spectrum bio- technology into my balanced, harmonized, resonant energy field. Since the inner self resides within the heart, that is where the work must begin. This is the initial spiritual energy center to unburden, balance and align to higher spirit. Unconditional Love is the essence behind all things in existence. This is the force that generates all things in creation. It connects us to the Absolute, Kronos, Helios, our Source, Creator, God, Allah, Yahweh, Jehovah, Brahma, Ra and so on, whatever you choose to call our Creator all these avenues lead to the same holy realm. This critical detail has been expressed in abundance in this literary work for a reason. Ultimately fact means that we must not judge others who follow proper ethical, good, loving forms of worship and organized faiths. Grains of truth have been planted within each of the common organized religions, Eastern traditions and western philosophies. It is our job to find these grains of truth on the road to building a spiritual belief system that resonates with our sense of self. This is the case for any adult person with cognitive, rational thought capabilities, sound mental faculties, free will along with a desire for truth, purpose, hope and faith.

Are you quite happy enough existing simply as a physical biological life form having a material experience? Would you be pleased to live such a limited form of existence until the day you die? Life will be exactly that if that is exactly what you want. On the other hand, do you search for light that you can't find? Are you looking for something spiritual that you cannot fully define? If this is the case, a mere physical existence is not enough for you. It is human nature to seek to know our Source and our purpose for being here. It is also a natural human curiosity to wonder how we got here. It is our human obligation to attempt to discover answers to these spiritual questions. Each generation is to make their contribution to this overall effort on behalf of the human race as a whole. As humans we are the detectives in our own mystery story. We are the only biological entities capable of cognitive thought and mental deduction. We are pre-equipped with all necessary elements required to transcend our conciousness during our biological experience. We are pre-programed to take interest in spiritual quests for light, wisdom and knowledge. We are internally programmed from birth onward to eventually develop a desire to seek out the Source from whence our souls emerged and will return to when our soul cycles of incarnations are completed. Inherently, we seek to know the source through worship, devotion, prayer, spiritual ceremonies, shamanic journeys, spiritual quests, personal ethics, religious study and so on. This is fundamentally because we are one with our Source, one with nature and one with all things in existence. All is One. Our seeking of light is quintessential human nature. The tools that we use on our spiritual journey can often be of tremendous aid in reaching states of enlightened awareness. The PMP 3.0 bio-field technology that we are currently reviewing, facilitated and quickly boosted this process into progressive action for me.

Initially, I began my energy healing work with brainwave binaurals and isochronics that target specific energy center frequencies within the heart chakra area. In the findings produced by Dameon Keller regarding solar spectrum calculations, these three chakra centers are composed of the following sympathetic frequencies. Our Schumann causal heart, positioned right below our astral heart resonates with a B note and 7.83Hz. This is the frequency of the Schumann Resonance. The theta brainwave range falls very close to this chakra frequency and as such, aligns with this causal heart. Second in place moving upwards is our well known astral heart chakra which is one of the 7 major chakras. Our heart chakra resonates with a C note and Alpha 1 brainwaves. Binaurals and Isochronics that utilize the frequencies of C256Hz and C128Hz are very beneficial to our mental, physical and spiritual health. The third chakra residing within the heart chakra center is a causal chakra that has been named the Infinity chakra in the research work of music producer, energy healer, Dameon Michael Keller in his book "Sounds Good, Sounds Great, Sounds Amazing". This energy center aligns our heart energy to much higher, more transcendent cosmic energies. It resonates with a D# note and Alpha 2 Brainwaves. This chakra also sympathetically aligns to the AUM frequency in the 12 note semi-tonal chromatic scale of notes. Infinity, as such, is an appropriate title to represent the energy of this second causal heart chakra. In my 1st book, *Awaken and Ascend* I provide information regarding chakras, solar spectrum frequencies, color and brainwave correlations. Calculations are based on reducing the solar spectrum Thz frequencies of color, down to audible Hz frequencies. This scientific, mathematical process of reducing inaudible light frequency to sound produces precise, accurate metaphysical results.

It is imperative to note that I exercised these three energy centers on a regular basis for 1 full year before discovering and adding Profound Meditation 3.0 Full Spectrum program to my energy healing routine. During this time, I also dedicated 1 hour a day to utilizing the Jose Silva Mind Control Program. These initial tools facilitated in establishing a spiritual connection between my biological earth vessel, my inner self and our Mother Earth. Our earth contains various elements, minerals, vitamins, nutrients, chemical compounds, essential principle materials, Prana, life force or Chi, electromagnetism and even consciousness. All these forces align to our human energy as we are of nature, from nature and our souls return to nature in between incarnations. Humans cannot exist in the absence of the Schumann frequency. This sacred harmonic frequency is the heart beat of our planet. We have an energy center within our biological body that connects us directly to it. As you have already learned, this is our Schumann causal heart which resides directly below our major heart chakra. As organisms that derive from nature we can access this force to either transmit or receive loving, healthy, mother earth energy. By sending back loving energy into our planet we can help raise the harmonic resonance of our earthly abode. This aids in healing our environment through the forces of unconditional love. The omnipresent force of love sympathetically generates a high frequency. Earth, much like ourselves is a living biological organism. Thus, we can also draw forth any type of healing energy that we require, directly from being connected to Gaia, our earth. When we align with the Schumann frequency through binaurals or isochronics or listen to theta frequencies, we automatically align to earth energy.

My heart began to open up to myself and show me the energies that needed to be cleansed very rapidly after I began using Profound Meditation Full Spectrum 3.0 biotechnology.

There are various energy healing tools available on the market which incorporate some very good modalities of energy work. When we discuss PMP 3.0 biotechnology, we are talking about an entirely different healing experience, with entirely different illuminating effects, that can change your life forever. I know this because this is the result it has produced for me. The difference between success and failure in regard to energy healing is having the right attitude and utilizing the proper tools. It is a fact that we are human biological lifeforms composed of earth elements. Consequently, our earth vessels cannot survive without our constant physical connection to the Schumann frequency. In this way, we are an essential part of Nature. Our material bodies are a literal biological bi-product of our earth. When astronauts go into space there is a device on board the space shuttle that resonates the Schumann frequency without which the astronauts could not survive. This detail alone should make it very clear to any intelligent, rational person, how very critically imperative it is that we take care of our earthly environment. Our Earth most certainly takes excellent care of us and gives us all we need for our survival. It is only fair that we do the same for her.

Our Earth Gaia, is only on loan to us as a race. This planet may be our home, but we don't own it and it is not ours to destroy. We have been granted no right to do that. We must all be aware of the fact that if our planet gets sick, so do the biological lifeforms that reside upon its surface. Do we not become ill when we drink contaminated water or breathe toxic air? Do our crops not fail and die if they are planted in contaminated soil? Yes, as these things are all connected. As divine biological manifestations we are one with each other and one with our planet. If earth dies, we die. All of us. It's that simple. Sounds like a crisis situation doesn't it? Do you think that the bulk of humanity is aware of this sacred connection? Do we care enough about this truth to make an effort help rehabilitate the damage we have done to our environment? By paying attention to the needs of our planet on a global scale we can all help preserve our living, biological home. Remove all negativity associated with lies or conspiracy theories. Step back and look at the big picture from a much wider perspective. Even if we do not agree with all the reasons behind today's global environmental protection policies, the core intention behind these plans, the essential ingredient that binds them all is the cause itself. The effort to save our planet and environment form any further decline or reckless devastation. Healing our planet is by no means a bad thing. We can ignore the politics if we disagree with them while still supporting the root cause behind the policies. Fighting against this part of the system only hurts our planet more. There are a whole host of excellent motivated reasons to radically improve our environment. Forcing humanity to care should not be a prerequisite. It should naturally be a privilege. There is a current fundamental need to focus on the health and conditions of this breathtaking planet without which we could not exist.

Each generation of humanity has the critical significant duty and obligation of leaving this planet in excellent condition form the next generation. Some of these people are our children. Some are grandchildren, nieces or nephews and their children who are to come in the future. We must think of these souls too. At the end of the day, we are all one family. Let us take responsibility for our living habits, industrial foot print and the detrimental commercial environmental impact that we have had on our only home in the last 2000 years. A healthy earth promotes excellent states of health and well being for all biological life forms on the planet. This applies to all living things from the plant and mineral kingdom, to the animal and mammal kingdom which includes the kingdom of man. It even applies to weather patterns here on earth. Everything is connected in our biological world.Nature is immaculate, intelligent and orderly by design.

After one year of focusing my healing work on my heart chakra center and earth connection, a sense of balance within body-mind-spirit was more often present and much more familiar. This was a tremendous achievement for me as I had lived with acute anxiety for my whole life up to this recent point in time. Ultimately, energy healing through sound was generating outstanding results. Through a use of binaurals I had achieved a high rate of noticeable, observable success along with a change in my energy to a higher state of vibration. At this stage a deeper connection to my heart chakra energy and holy guardian angel within, the fundamental essence of the inner self, was evolving and developing a more rapidly. It was at this stage of development that I upgraded my healing program to incorporate daily use of IAWAKE , PMP 3.0 Full Spectrum Meditation bio-field technology.

At this critical juncture it is imperative to note that the method being described to you is the exact manner in which I conducted my own personal energy healing practice. This is the path that I took, but it is vital to point out that PMP 3.0 is a Full Spectrum Program that attunes all brainwave frequencies and chakra energy centers quite successfully on its own. Adding other programs to your bio-field technology healing work is entirely unnecessary. This system can produce powerful, magnificent, highly successful results on its own. That being said, use of this programming will not interfere with any other type of healing work that you are conducting. The PMP 3.0 bio-technology system serves only to enhance any other energy healing work or technique that you commonly practice. Essentially, in order to generate tremendous results and benefit greatly from IAWAKE bio-field technology you do not have to take the exact same route that I did. The path I chose was based on my knowledge of visible solar spectrum light correlations to our human brainwaves, biological and sympathetic bodies. I applied my knowledge to creating a program that was built to re-align my emotional self and physical self in perfect balance and harmonic frequency. My initial goal was to heal my emotional condition which consisted of acute anxiety and it's associated chronic nausea. Proper use of binaurals and isochronics were successful in helping me to achieve a more balanced state of wellness. The vehicle that raised my consciousness to lofty expanded heights allowing for spiritual transformation, expanded awareness and mental transcendence was IAWAKE biotechnology. Even if working with sympathetic frequencies and bio-field technology is new to you, this program is custom designed to be powerful enough to produce noticeable, beneficial results. No previous experience with meditation or energy healing practice is required.

Profound Meditation3.0 can be found on the official IAWAKE website www.iawaketechnologies.com . This multi disc bio-technology program consists of tracks that help with energy releasing and sessions that promote transcended states of consciousness. Both fundamental energies are packaged together in one perfectly ordered, highly advanced audio mediation system. This product sells for $227.00 u.s.d. For some people, this price may fall into a steep range. If you live in a reality, where your budget requires you to save money in order to buy yourself a random gift, then this program gives you the most perfect thing to save up for. How much money is expanded awareness, transcendent consciousness, a heart filled with internal peace, unconditional love and a connection to inner self, worth to you? To me, gifts such as these are priceless fruits of the spirit. No monetary value too great, can be placed upon these assets. The spiritual rewards that I have received from this program and the results that I have attained in regard to expanded consciousness utilizing this system, have paid it's dollar value in returns 10 times over. The effects of this program regularly contribute to my ability to place my thoughts and mental patterns in a positive vibrational state. The ability to instantly do this comes about much more naturally, with much greater control and ease than ever before. This meditation system is a binaural bio-field technology audio program that ranks #1 on the market for anyone wishing to re-invent themselves into a more productive, spiritual, top quality, peak performing, highly effective 3.0 version. First of all, it is significant to note that not all systems are created equal. Profound Meditation 3.0 Full Spectrum is not your average system. This bio-technology crowns the list of sound healing programs because it is completely different from all to others by way of construct. After 1 year of using this system, I have developed better control over my mental and emotional faculties than ever before.

Much like all other audio entrainment meditations or frequency energy healing systems, there are safety regulations and health cautions that must be taken into consideration prior to use. For example, pregnant women or people with pacemakers, a history if tinnitus, seizures and so on are not recommended to use binaural or isochronic programming. It's excellent advice to go to the IAWAKE website and read their introductory material which includes their legal disclaimer and a list of health cautions. This material outlines all required safety precautions related to using this amazing consciousness expanding resonant bio-field technology. So before you even purchase this mind blowing system update yourself on these particulars. In most cases, these precautions only apply to a small number of people whereas the majority of the population will likely have no issues using binaural beats or bio-technology. In these cases, this form of energy healing has zero negative side effects and endless infinite benefits. When in doubt, consult your doctor before use if you have any concerns regarding healthissues or any pre-existing medical conditions.

As noted, Profound Meditation 3.0 Full Spectrum is designed

in a completely different manner than the common forms of binaurals, isochronics and brainwave entrainment tracks that we are used to. This unique system consists of very specific, carefully designed frequencies that provide long lasting healing benefits. You will see results not very long after incorporating this program into your daily healing routines. This system is built to generate very deep states of meditation at a fast pace. In contrast, other traditional techniques, systems and programs offered on today's market can possibly take years to master. This biotechnology boosts those achievements creating the same results at a much faster pace. All that is needed is just this one program. Do you wish to expand your awareness? Do you wish to elevate your mind to the greatest levels of consciousness that you can potentially achieve? Did you know that when you excel to your most efficient level of intellectual, and emotional potential that you become a vital asset to humanity ? In our most prime, optimum state we are capable of contributing selfless service to our race. This is so. Ultimately this is what we are here for. What we do to the one, we do to the all because everything alive is connected. We are one.

Additionally, our individual spiritual progress literally makes a divine contribution to raising the overall consciousness level of the human race as a whole. We all have a responsibility to do our part to fulfill this spiritual requirement. Achieving a state of expanded consciousness promotes human evolution and can potentially initiate scientific and spiritual revolutions. Every one of us has a service to render to society and to contribute to the human race as a whole. What do you have to offer? What is your talent, craft, hobby or trade skill? What is your superpower? We all have something to offer humanity. It doesn't X-ray eyes and fantastic biohazard mutations to be a super hero. Batman is a stellar example of wisdom, ingenuity, vast intellect and sublime immaculate human ethics. He was always calm in the heat of battle. He was a genius with impeccably extensive rare knowledge. His tremendous wealth gave him access to all the tools and connections he ever needed to save the day. Batman demonstrated super human strength. He was a master of the mind and human conduct. In countless ways, Batman is a good lead to follow.

Our inner self holds the key to what it is that we have to offer humanity as an gift of selfless service. This aspect of ourselves already knows this answer the moment we are born into this world. Thus, connecting to inner self, helps us to discover and demystify our spiritual earthly mission and unravel the divine purpose of our own existence. The heart is where this truth and these answers can be found. It is our perfect, spiritual inner self who is inherently connected to the one collective mind that knows all things. Some refer to logging into this vast information field as accessing the Akashic Records. Some call it tapping into the energy grid while others may refer to it as channeling spirit. Is this a goal that you seek to accomplish? Do you wish to reach this level of consciousness in the least amount of time possible? If so, then this bio-field technology program is the best tool to help to you generate this state of awareness and existence.

In regard to progressive development, there's a lot more to IAWAKE than an independent experience. Our purchase of this system includes ongoing education offered through an opportunity to become an integral part of the Global IAWAKE Community. Fundamentally, each of us can become a member of a spiritually ascended family focused on growth, oneness, love, enlightenment, compassion and evolution. The IAWAKE family has members from all around the world who share healing experiences, goals, common interests, tips and stories. Becoming a member of this group can offer further valuable guidance along your path to enlightenment while connecting you to other people who are spiritually empowered and awake. The excitement goes on and on.

In addition to PMP 3.0 Full Spectrum, IAWAKE offers a vast number of programs that are designed and tailored to suit a whole variety of specific healing goals. All you need to do, is log into the website and tap into this spiritual network. The members of the IAWAKE community engage in interconnected global meditations on a regular basis at key times throughout the month. All members are invited to participate if desired. By joining in on these meditations, spiritual minds can unite with the common goal of contributing to raising the conciousness levels and harmonic resonance of our race as whole. That is very powerful output. Do you wish to be part of this great healing force? This bio-field meditation program doesn't just give you the tools to transcend consciousness, it gives you a whole community of like minded people to evolve with, if you so desire.

Everyone can inherently recognize ascensional energy when they feel it. The new age movement is happening. Spiritual awakenings are happening. Modern science has yet to show us an origin of creation model that corresponds with the repetitive known processes of Nature, metaphysical science and Universal Law. If this event occurs, a new age of light will instantly be upon us. In this light there will be no denial of oneness. In this light there will be proof to establish the one thing that we have pondered all along, the chance that we are not alone in the universe. We live in a universe constructed upon endless repetitive patterns of creation and manifestation. We can metaphysically prove reflective patterns of existence. Bearing this in mind its pretty safe to assume that our entire race is also reflected out there somewhere else in the cosmos. In fact NASA has already discovered the existence of other earth like planets that may contain water and an atmosphere capable of supporting biological life. For our earth and our race to be the only one of its kind, defies the Natural Laws of Science. I consider this notion to be very logical, rational, mysterious and exhilarating all at once. Don't you? What are your thoughts on the subject when all things are considered? It is quite a profound, intriguing possibility to ponder.

Profound Meditation 3.0 biotechnology commits to its impeccable reputation. It actually works exactly the way John and Pam Dupuy intelligently designed it to. That being said, there is no form of energy healing tool on the market that can do the inner work for us. No tool on the market can substitute for directed will and positive intent. When it comes to the inner work, the healer is you. Spiritual teachers, gurus and shamen can teach us the necessary knowledge required to live a good life in accordance with Nature and Universal Law. However, internal healing has to be chosen of our own free will and committed to individually. We must willingly dedicate ourselves to our goal of healing in order to achieve the best possible results. In this way, the inner self can be much more easily accessed. This is the ultimate result that we are essentially attempting to generate. A dramatic shift of focus, perspective and paradigms occurs when we operate from the vantage point of the inner self residing within our heart. Connecting to inner self energy contributes to mental expansion and broadening of awareness. This sacred alignment promotes advanced levels of spiritual connectedness to all things in Nature and to our Source itself on an internal level. The process of turning inward to align to inner self is heavily facilitated by the use of the IAWAKE biotechnology system. The Profound Meditation 3.0 Full Spectrum system enhances our emotional energy. Continuous use improves our general state of mind and spiritual appreciation for simple presence. This programming helped me learn how to appreciate the sound of silence. It gave me the internal patience to simply be still enough to listen, hear and feel the energy of the spirit forces of Nature that are moving all around us at all times. This bio-technology system eventually gave me the ability to learn how to slow down, recalibrate my energy whenever necessary and just breathe.

Ultimately, my state of mind shifted away from worrying about greater influences outside myself over which I have no control and cannot change. I then became more focused on that which I can change or improve about my life within my own personal circle of influence. This encompassed the immediate aspects of my personal life, the events, circumstances and the people in it. In this healthier mindset, I was able to eliminate excess stress, worry or panic over circumstances which I have no power to alter. Daily use of this powerful biotechnology system sharpened my thought processing abilities and strengthened my rational sense of logic and reason. This in turn, aided me in calmly finding my place within the confines of a constantly changing environment. I literally learned how to become the chameleon and continuously adapt to influences in the space around me as opposed to reacting. Reacting was the old programming that PMP 3.0 gave me the ability to overpower. In this way I was able to reduce mental stress, learn what is important to me and mentally grasp the understanding of that which I can and cannot change. Empowering thought processes such as this contribute to a healthy, balanced mind, body and spirit rooted in logic and rationality.

Exposure to this bio-field technology has an amazing effect on my thought processing abilities. Consequently, it is now much easier to shift my thought forms from a negative to a positive headspace with much greater ease and control than I previously ever had. In the next chapter, I share a simple technique for instant brainwave and heart rate entrainment. The functionality of this described method, is heavily boosted and empowered once one establishes the mental states of awareness that are generated by PMP 3.0 Full Spectrum. In short, my brain works way better, is much sharper and faster at mental deduction as well as memory retrieval. Using this technology, I was able to eliminate parts of myself that fight against rationality, logic and reason. All that remains after a mental house cleaning such as this is the desire to become the best person that you can possibly be. It is our duty to discover gifts that we internally house. From these inherant natural talents, come selfless contributions that can make the world a better place and spiritually uplift our race as a whole. Priceless fruits of the spirit can more easily fully develop once a mental and emotional release has been conducted.

The release tools embedded in this highly advanced biotechnology can be used anytime we feel negative energy affecting our resonant bio-field. When this occurs we simply need to re-calibrate our energy centers and brainwaves to a higher frequency. Further more, enhancing my spiritual awareness and consciousness using PMP 3.0 powerfully facilitated my ability to establish contact with and align to my inner self. Once I achieved this goal, the internal relationship with this force and continued use of IAWAKE biotechnology promoted my ability to connect the collective consciousness. When this occurs, you will see light. Externally or internally, each experience will differ. Keep eyes open nonetheless. Silver light is used for angelic transmissions. Gold light is used by ascended masters and saints. There is absolutely nothing evil out there. Anything evil resides within the hearts and minds of man. The only demons you will ever encounter in altered sates of consciousness belong to you. They already live inside of you. These are the negative energies, built up over the course of lifetimes that must be released. This transcendental mental state, once generated makes contact with higher forms of spirit possible during channeling work. It also gives us an internal sense of knowing that a higher spiritual presence is simply always there, all the time. Immortal conciousness exists in Nature everywhere all around us.

Every biological material organism has its own bio-field. It emits this energetic field sympathetically in the form of electromagnetism. At the highest levels of consciousness we can align with the collective mind. Through the collective consciousness, we gain access to the realm of archangels, saints and ascended masters of the human race. This is exactly what channelers of spirit and mediums are tapping into when they make contact with other forms of spirit. With determination and persistence communication can become possible but very few people can attain this level of mental transcendence. Not because it is impossible. More so because not everyone has time for a daily 1 hour commitment to acoustic energy attunement. Repeated daily use of Profound Meditation 3.0 was an immaculate aide in helping to establish the connection to inner self that is required to access a transcended level of consciousness. Furthermore, once you can commune with your inner self effortlessly, the gateway to higher spirit realms of communication can now potentially open. Commitment must be combined with intent, directed will and a strong desire to achieve a state of ascended consciousness. In this mindset daily use of this biotechnology program is the ultimate key to open the gates to the greater forces that exist in Nature all around us. I will gladly stake my name on that fact. The collective consciousness and the energy that can be drawn forth from it exists within us, extends to the space all around us, beyond our current location and all the way into the realm of the Absolute. The level of collective consciousness that we can reach by way of practice depends on how successful we are in our own personal energy healing exercises as well as the amount of time and work we are willing to put into it. Nothing worth having comes easy. The doors behind which we find the most brilliant light are the heaviest doors to open. But once the gate is cleared, the immense amount of effort pays off 10 fold. Personal improvement and the things we work the hardest for in life to accomplish often turn out to be our greatest achievements. This information is key.

Profound Meditation 3.0 is designed to initially generate a complete energetic detox. The cleverly constructed session tracks cause a great release of outmoded, unwanted energy. This biotechnology utilizes the proper harmonics to promote over all healing and the bio-field resonance required to align our hearts to the unlimited unconditional love found everywhere in Nature. From the core of our planet to the stars above us, the spirit forces of Nature are omnipresent. These forces are a part of our divine essence. PMP 3.0 Full Spectrum is a powerful, versatile meditation tool. If you like the effects that you achieve using this system, you can use it for years to come without a need for any other energy healing system. Because of its unique design this program will continue to evolve our spiritual levels, mental awareness, consciousness and sympathetic, biological vibrations for as long as we choose to use it. The extended promise of continuous development to new heights of spiritual evolution was a key factor that gained my interest and desire to test this system out. Biotechnology engineered at this advanced level is a one time purchase that meets an endless array of physical and spiritual healing requirements that can be utilized for life. It's literally a gift that keeps on giving.

This program was engineered using the latest most advanced bio-field technology that assists in personal growth and development, intellectual evolution and spiritual awakenings. The added potential for continuous growth and development can be seen, felt, experienced and recognized within my resonant bio-field. These changes began to take effect not long after I began using this biotechnology on a daily basis. It is imperative to note, that for each person the experience of awakening will be different and may take longer or less time than it does for others. Ultimately the rate at which your consciousness develops depends on the current level of spiritual awareness that you have already established prior to initiating this program. It is of critical relevance that we start with the best energy healing tools available if we wish to achieve the most optimum results. If you are a regular practioner of energy healing work, sound therapy or meditation with prior success, this system may work very rapidly for you. PMP 3.0 promotes a complete reprogramming of mental and emotional faculties. This biotechnology does hold true to that bonus feature. This is exactly what it has done for me within the time frame of one year. There were several levels of obvious noticeable development that presented in stages of growth with continued use. People around me have witnessed a change for the better in my overall personality. My anxiety condition has been reduced to nil. When my emotions are rattled, I can much more easily take charge of that energetic imbalance right on the spot and choose to remain unaffected. There will always be stimuli in life to react to. How we choose to react to any type of circumstance or situation is what makes the difference.

Various stages of noticeable development generated new levels of consciousness all throughout my whole initial year of using this amazing IAWAKE program. I experienced this aspect as noticeable progressive stages of mental expansion and broader thinking. This development was coupled with the shedding of outdated thoughts, beliefs, habits, patterns of behavior and so on. The resulting effect was an active, continuous, ongoing correction and fine tuning of conduct as opposed to acting out emotionally in ways that I once did. New more productive modes of thought and patterns of behavior began to replace limited ways of thinking. Very little time was required for me to feel and recognize obvious changes in my thought processes, energy levels and mental awareness. These changes began to formulate and take effect after only a brief period of daily exposure to PMP 3.0 Full Spectrum.

As previously stated, the rate of speed at which you begin to see results such as these varies for each person. Experienced energy healing practitioners will logically achieve faster results than a novice who is experimenting with meditation for the first time ever. That being said, through extended use your mind, body and spirit naturally begin to attune to a higher vibrational frequency and align with higher states of consciousness. This energy promotes balance of body, mind and spirit while generating internal self healing abilities. This is absolutely one of the best energy healing tools available to us and the only one of its kind. This advanced energy healing system promotes more rapid access to different levels of my intellect, increases my creative potential and has rocket launched my physical productivity rates to the moon and back. The heart centered connection and mind expanding effects that these audio tracks produce are of critical fundamental relevance. These states of being are key factors in generating the powerful connection that I have gradually established with my inner self.

In my self healing energy work I am now practicing my ability to activate different brainwave states at will currently within the ranges of beta and alpha.I am also exercising my ability to gain greater con trol over the separate hemispheres of my brain by attempting to entrain my brainwaves within these frequency ranges. My ability to manage a shift between these brainwave states and to acknowledge this alteration occurring quite naturally on its own throughout the day is increasing with continued practice. Mental abilities such as this arise through repeated use of this magnificent programming. PMP 3.0 biotechnology worked immediately on my well attuned energy centers which had been calibrated to prime harmonics for a full year prior to exposure to this system. Exposure to this bio-field technology launched my spiritual evolution into action with increased mental awareness. In effect, a profound rapid connection to inner self and then to the unified mind residing in the vector field was established. My intellectual capacity has also increased along with my mental processing abilities. Surprisingly, even the rate of speed at which solutions to complex issues can be produced, acknowledged, processed and applied has been magnified. The rate of speed at which you develop expanded awareness will vary depending on your level of commitment to spiritual evolution. That being said, the benefits of these audio sessions, when combined with meditation and creative visualization, are beyond astounding.

Now, on a daily basis I move ever further away from a mindset that allows ego to interfere with spiritual growth as I progress along in my healing journey. My mental capacity has expanded to macrocosmic levels, wide scope broad range forms of thinking and deep contemplation of grand, magnificent concepts. This mindset has drawn me closer to divinity both within and without. Furthermore it has generated a great desire to study origin of existence theories, of which I extensively spoke in the first book of this series, *Macrocosmic Reflections in a Microcosmic Mirror*. Contemplation of grand, profound concepts such as galaxy formation, the holographic universe or the purpose of our existence is a bi-product of working with bio-field technology. This program offered to expand my awareness and it has indeed done that. Once our heart chakra is activated regularly through energy heightening tools such as this, a powerful attunement to higher vibrations and to inner self can be established. This connection promotes an ongoing state of existence in the light of an awakened being empowered by unconditional love. In Eastern philosophy this awakened, attuned sense of consciousness and ascended sense of being is often referred to as a state of Nirvana or Samadhi. Its not very complicated once demystified. All we are doing is living in a higher vibratory state of existence. In the western world we see this connectedness as living in oneness with Nature and all other beings in a state of compassion and unconditional love. In this mindset we are empty of all lower aspects of ego self and fully aligned to Universal Law much of the time. Either way we are referring to the same quintessential, omnipresent energy that is experienced at various ascensional levels of spiritual development.

It is imperative to remind you that reaching states of being such as this requires willing participation to do the inner work and devout desire to connect to the collective consciousness. In addition, a valiant effort to practice some form of meditation or creative visualization is the asset that boosts the potential of bio-field technology in the right direction. Success in any healing work is determined by our level of personal devotion, daily commitment, will, faith, effort, desire, directed intent, dedication, effort and desire to progressively evolve spiritually. Following this same basic, fundamental application for achieving any desired level of ascended being is a requirement for any type of spiritual tradition or healing system here on earth. In addition, what I have also discovered, is that the more knowledge I acquire, the more information the ascended masters of the collective mind can impart to me. As I develop a deeper comprehension of the scientific processes of Nature and the interconnectedness of all things in existence, the more attuned I become to the many omnipresent spirit forces found all throughout the biological world we live in. My perspective has expanded and in my research, I developed an interest in studying quantum-biology from macrocosmic to the microcosmic levels of creation. This type of elevated frequency and widened, expanded perception are intense, highly enlightening forces for anyone to experience. Tremendous amounts of academic and creative potential, are magnified and flow forth effortlessly once channels such as these are consciously accessed, mentally activated and opened. The backbone of this force is the unconditional loving essence and consciousness of our

Creator. We must choose to align with this force of our own free will through directed intent and inner absolution.

Profound Meditation 3.0 Full Spectrum comes with an extensive manual explaining how to properly use this highly advanced system of bio-field technology and harmonic resonance. This program works producing optimum effects for beginners, practitioners at intermediate levels and more advanced users. It is for everyone in a sense that it has been intelligently designed for any practioner of meditation regardless of skill level, training level, experience or lack thereof. The effects of this program can be felt within our resonant bio-field. As such, you will know after only a short time of exposure to the session tracks, which ones work best for you and in what combination to use them. Everyone experiences the effects of meditation differently. As such, the suggestions made in the manual are meant only as guide. Certain tiers that they suggest may work better for you than others. Ultimately it is up to you to determine which audio tracks affect you the most profoundly then mix and match the arrangement in any order in which way you wish to use them. This system allows the user to align their own energy to the powerful bio-field resonance in a tailor made format based on results desired, experienced and achieved. Profound Meditation 3.0 Full Spectrum, exercises your brain and nervous system simultaneously. This bio-technology energizes our chakras and contributes to balancing and harmonizing these critical energy centers.

Training and exercising our brain, mind and awareness is necessary if we wish to expand our consciousness to a level of development required to generate the state of existence wherein we have mastered the Ego Self. Adding inner work such as creative visualization coupled with meditation during the IAWAKE audio session amplifies the software's potential. In turn, this maximizes and manifests even greater, more positive results. Many states of expanded mind and elevated spirit can be progressively accessed using this system. Benefits of this type of mental and energetic bio-field re-alignment and system upgrade include increase in compassion, expressions of forgiveness, feelings of unity, oneness and interconnectedness to all things. It promotes alertness, attentiveness, patience, peacefulness and loving energy. In addition, it creates a mindset that operates in the present moment. We begin to exist with a conscious awareness of our basic presence and the silent witness within us. Energy healing, consciousness expanding tools such as this induce higher states of biological and spiritual vibrational existence. To reach the most advanced heights possible, a mental and emotional house cleaning will always be a fundamental requirement. This vital part of the ascension process is a critical step without which successful mental transcendence could never be achieved. This implies our ability to let go of and release every single thing that we see as a defect or an unnecessary piece of mental baggage. Unhealthy beliefs and useless mental paradigms get thrown to the wayside. Obstacles such as these are the anchors that pin us down whenever any type of spiritual ascension is attempted. One must confront their fears, inner psychological demons, choose to release negative energy and let them go. These things serve you no justice on the path to righteous light. The many loving Spirit forces of Nature commit to guiding every human soul who makes the honest attempt to heal themselves. We all have the option to make the world a better place by personally committing to conscious transcendence and spiritual evolution. When we each choose to do so, it raises the vibration of our species as a whole. That's how nature operates. This is one of the many ways that our Source helps humanity create order out of chaos on our physical, material plane. We are all vital components of nature.

Continued use of the energy releasing tracks inevitably caused negative energies to arise and begin to dissolve until it could no longer be felt as a dominant force within me. Unhealthy vices and bad energy arose, moved outward and passed through me then out of my energy field, followed by detachment. This process was a repeated occurrence that continuously lightened the load along the way until a peaceful empty space was created where obstacles and mental vices once resided. With continued use of PMP 3.0, this release process is generated, guided and conducted so as to avoid shocking our nervous system or causing energetic mental overload. As the releasing tracks go about doing their amazing work, the load gets lighter and lighter all the time. We shed the darker weaknesses, fears and energies that tend to weigh us down in life even when we are unaware of it. Faults and weaknesses tend to get in our way and limit our spiritual, emotional, mental and intellectual growth as well as our success in life. Our weaknesses can legitimately be referred to as our inner demons. They should not be feared, coveted, worshipped or catered to. You can hang on to your ghosts, cling to your familiar faults, defects or vices and keep your closet packed to the hilt with dusty old skeletons of the past, we all get to choose. But bear one vital detail in mind. These hypothetical possessions only promote emotional bankruptcy and paralyze spiritual growth on a colossal scale. Letting go is a much better option. Releasing requires direct confrontation of common human character defects. These can include bad habits, weaknesses, fears, limitations, emotional vices, addictive behavior or any other negative energy. After discovering this disempowering energy, the next critical step is to label the unwanted vice as outmoded, outdated, unproductive and useless, then let it go. Only in this way can we fully cleanse our bio-field, aura, mind, heart and energy centers and eliminate old, expired programming. Vices are nothing more than obstacles on the path to light. The solutions to many problems we face in life will always be discovered at the roots of an issue. Perceiving any obstacle from this vantage point can help us eliminate the issue and tap the roots of the problem simultaneously.

Releasing and letting go is an individual decision and a personal choice. That being said you will never reach a lofty height of spiritual ascension until this important work has been conducted and concluded. Unloading unnecessary baggage facilitates a much faster, more efficient arrival at transcendent destinations. Critical acknowledgment of this point factors into spiritual evolution on a mandatory level. It has been stressed in this literary work numerous times because of its tremendous significance. Assisting in achieving this state of existence is exactly what Profound 3.0 IAWAKE technology was designed to do. The bad must come out and be sanitized away before high vibrational positive energy, such as the unconditional love of our Creator and the spirit forces of Nature can enter our auric bio-field upon invitation. Once we reach such a spiritual state of being it is imperative that we continue our consciousness development exercises to even further enhance and shape our spiritual development. Continued practice of this meditation system eventually gave me the ability to enter into elevated states of consciousness more effortlessly. As my spiritual development progresses, the higher vibrational frequency that is generated is becoming a more natural state of being. This effect greatly empowers any other learned spiritual methods and practices that you may conduct such as astral projection techniques.

Our deepest potential is to connect to our Inner Self and remain aligned to this great force of Spirit within us. When we achieve this bond we begin operating from our heart center. This is a whole new state of mind and presence. Meditation and creative visualization during the session speeds up the effect of this consciousness expanding program just as it does with any other form of energy healing. Without incorporating inner work into your PMP 3.0 session, this technology will still work. It will still produce amazing health benefits and the intended effects will still manifest just not as powerfully as they would when used in combination with inner work. Inner work implies meditation, creative visualization and slow and deep rhythmic breathing. This biotechnology aids tremendously in achieving mental states essential to becoming a master over the objective outer self and it's many vices.

Profound Meditation 3.0 Full Spectrum was uniquely constructed to activate both hemispheres of the brain as opposed to just focusing on only one part of the brain. This feature alone makes it magnificent beyond measure. This system targets the upper right lobe of our brain which results in increased overall coherence, cognitive thought processes and mental capacity. It also stimulates the upper left cerebral hemisphere of our brain. Exercising both areas of the brain at once generates a whole variety of positive results and activates the full range of brainwave frequencies. This system promotes and increases mental, academic, spiritual, intellectual, and consciousness development all at once. Accessing our dual hemispheres during energy work intensely magnifies any results received from stimulating and targeting either hemisphere or cerebral lobe on its own. Gaining any amount of mastery over moving in and out of certain brainwave states at will using the power of your own mind is a highly useful skill to acquire. Catching brainwave states in active engagement and slowing them down to facilitate information processing is another amazing skill that I have been practicing. I exercise this skill on a regular basis as it is one of my own personal keys to attaining a balanced mental and emotional state. I am learning to slow down the mental wheels of a mind that occasionally enters academic overdrive. Altering consciousness, heart rate and entraining our brainwaves at will requires concentration and memory skills. What I can truthfully tell you, is that the Profound 3.0 IAWAKE system does calibrate and upgrade our bio-field energy so that we can develop fundamental spiritual skills such as this.

In order to generate the spiritual awakening we all seek, repeated exposure to this bio-field technology is critically essential. This meditation tool is like no other. This uniquely designed system has produced the most profoundly enhanced levels of expanded awareness and elevated consciousness that I have ever experienced. PMP 3.0 will help you generate the most optimum results that you are personally capable of attaining through bio-field technology energy healing. Persistent use and the regular dual hemisphere activation are key to the immense power encoded within this system. Your experience may not be identical to mine, but I can tell you that with repeat use you will absolutely see results and changes for the better. You can witness this in your overall thought processes, attitude and demeanor. In any case, my example is one of incredible success developing an intense level of cosmic attunement. States of consciousness of this magnitude are only attained by exerting 100% into the mission of spiritual evolution. This program gave me what it initially offered, a profound spiritual awakening. One that has helped me systematically develop a whole new way of thinking and a whole new perspective of things in my own environment. I have reached a whole new level of heightened awareness and expanded consciousness. I have established deeper contact with my inner self. I have formed a new perspective on life. Most significantly of all, I have wiped my emotional slate clean and unburdened a lot of old, unnecessary mental clutter. As I continue to work with this system, my ability to control my brainwave states through the simple methods that will be discussed in the next chapter has been greatly enhanced. It continues to become easier as a result of establishing a deeper, sharper awareness through ongoing exposure to PMP 3.0 bio-field technology.

In my own experience, all aspects of my energy healing

program contributed to the level of ascended consciousness that I have established. PMP 3.0 Full Spectrum was the main, critical tool that helped me to align to higher levels of consciousness. That being said, all other aspects of my energy healing program played vital roles as well. These other tools also aided in preparing my mind and my biological vessel by harmonically attuning my chakras. This did contribute to the massive boost that PMP 3.0 generated. Every part of my program was a factor that promoted my developing ability to consciously and willingly connect to source energy. That being said, it is this biotechnology program that contains the required energetic elements that propel our ability to align to the collective. Ultimately, this system greatly enhances the results of any other frequency tools that I commonly use. This result is exactly what the designers of this brainwave entrainment system hoped to offer.

All technology used to create PMP 3.0 was heavily researched and intelligently formatted. The system was designed in a uniquely structured manner. Highly advanced, powerful bio-field technology has been incorporated and encoded within this system. IAWAKE technology is original and the only system of its kind. In all honesty I do still use binaurals and isochronics of the solar spectrum frequencies mentioned in my 1st book *Awaken and Ascend.* These tools are still useful for specific energy center calibrations or grounding needs. However, none of these tools were designed to specifically transcend your conscious and produce a spiritual awakening in the same way as bio-field technology does. Nothing else on the market can generate the same results. Its pretty impressive to consider that this programming will literally produce even more transformative effects on my sense of being with continued use. As such, it will be part of my energy healing work for years to come.

What causes this scientific system to stand put so boldly from other audio entrainment programs is the immense amount of structuring and layering of multi sympathetic carrier waves and modulating frequencies. The bio-field technology encoded into this system pushes the physical and sympathetic nervous system in a unique specialized manner that is totally different from methods used by all other systems. With regular and repeat use, I quickly noticed positive changes in my overall energy levels. My sleep was more restful, my general mood was elevated to higher vibratory states, my attention span became sharper and more concentrated. My energy slowed down to a healthier more relaxed pace. From this point forward, a sense of inner peace, compassion and global unity began to emerge. The desire to heal the environment, save the earth and great number of other spiritual impulses arose. Higher productivity rates and a sense of internal connectedness have literally now become more common place. These and so many other positive results have been experienced in my own personal healing process using PMP 3.0. It is indeed, very profound. As promised.

PMP 3.0 uses key signature multilayered bio-field technology. It is comprised of various audio entrainment techniques, psycho acoustic principles and bio-field frequencies. The sympathetic harmonics are directly tuned into the harmonic resonance of our chakra energy centers. This system targets the full spectrum of brainwave functions and states in a specialized manner. What is being described here is nothing less than a highly advanced, extremely functional neuro-technology energy healing system. This is one of the greatest treasures that I have discovered in my search for light, wisdom and illumination. This system has been a magnificent aide for initiating, guiding and successfully conducting my transcendent transformation of consciousness. It further promoted, empowered and produced all necessary elements required for spiritual evolution. In this higher vibratory frequency I have reached new levels of consciousness while moving through various states of transformation. Thisis the manner in which we reach new levels of ascended consciousness that can become a common state of mind. IAWAKE bio-field technology is the gold that generates that positive transformation faster than any other entrainment tool that exists on todays global market.

Once we connect to deeper more spiritual aspects of ourselves we have the potential to access these energies at will. PMP 3.0 is a powerful system that incorporates Delta frequencies with low modulating Epsilon frequency waves as well as Gamma 1 and Gamma 2 hyper gamma brain wave states. Very transcendent states of consciousness can be developed through various arrangements and combinations of these frequency entrainment ranges. As promised, PMP 3.0 produced states of enhanced cognitive thought processes, sharper mental precision along with keen visual perception. All functions of my mental faculties were effectively upgraded. I gained effortless access to the inner self. Energy levels elevated and   productivity increased. Furthermore this product expanded my creative potential beyond measure. These are some of the many gifts and rewards of investing in this product on a physical, emotional and spiritual level. I have much gratitude for the IAWAKE team for creating this immaculate energy healing medicine. *They have given a priceless gift to humanity.* I have never met John or Pam Dupuy, I am a random customer who purchased their product understands its unbelievable value to the world of energy healing practices. Otherwise I am not affiliated with them. What they scientifically created in their bio-field technology sound studio is a masterpiece. It is the perfect pill to swallow for those who wish to establish the ability to tap into the energetic matrix of the higher spirit realm. As a personal testimony the fact that this program review deserved a whole chapter in this book clearly indicates its ultimate quintessential value to me in regard to spiritual development.

At this pivotal juncture it is imperative to inform you that I did not use PMP 3.0 on its own. I chose to incorporate a second piece of IAWAKE bio-field technology into the mix. Essentially, I would be leaving out some very important aspects of my spiritual evolution if I failed to mention this. There was a very significant, relevant counterpart that I utilized along with this system for the first full year of practice. Together they were the great powerful, golden key to my personal transcendence of consciousness experience. In other words, my ability to connect to inner self and to spirit energies outside of myself. IAWAKE has a massive selection of biotechnology tools, audio tracks and meditation sessions to cover a wide range of spiritual and physical needs. I personally feel that the exact particular combination of bio-field technology that I carefully chose, provided me with the maximum ability to scientifically open the golden gateway to the higher realms of Spirit. That is what I accomplished and exactly what I had originally set out to do. So what is this second immaculate tool which is most certainly essential in developing the ability to activate the Golden light of the realm of saints and masters? The second bio-field technology system that I chose to utilize can be purchased on the IAWAKE website for a retail price of $57.00. That is not a lot to ask for an entrainment tool that actually does stimulate and activate our crown and super conscious chakras with regular routine use. Infinity Lambda Brainwave Meditation is the second system that I chose to alternate with Profound 3.0 Full Spectrum. In order to achieve the fully ascended state that I was seeking, I incorporated this second tool into my healing program. It was a personal choice, not a requirement. For as I've already mentioned, PMP 3.0 is a Full Spectrum entrainment system that was designed to work on its own. In short, no other tools are necessary. My goal was to propel my consciousness to the highest levels possible in the least amount of time using audio biotechnology. Does this program work? Does it deliver the higher levels of transcendence that it has been designed to generate? In my experience yes. Its worth every penny.

Infinity Lambda Brainwave Meditation was intelligently engineered to deliver what it promises. Any amount of effort I placed into this program paid off ten times over. Just as its immaculate counterpart, Profound Meditation 3.0 has done. The Lambda brainwave frequency stimulates the brainwaves of our ultraviolet super conscious causal crown chakra. This energy center is located right above our head and crown chakra. Within this energy center is a powerful connection to the spirit forces of Nature at higher vibratory levels of consciousness. Activation of this powerful energy center facilitates connection with the all knowing collective consciousness, known to some as the Akashic field. In combination, these two systems of bio-filed technology powerfully exercise both causal energy centers that must be activated in order to transcend your consciousness. Seekers of light, follow this advice. Commit to this program and I promise you, results will come. For me, this combination was like high voltage to my biological system. In a radiant, resplendent way. This revolutionary brainwave technology that I used has become an integral part of my daily life. It has offered an granted me the ability to re-invent myself into a much better model of mind ,body and soul. In a higher state of existence I can be of muchgreater service to humanity. As each of us should be. Infinity Brainwave Lambda Meditation takes your brainwave state down through the very low undulating frequency of Epsilon all the way up to the consciousness expanding brainwave states of Lambda. The Lambda brainwave state generates and empowers feelings of wholeness and oneness. It facilitates spiritual transcendence and generates deep meditative states. It amplifies feelings of bliss, joy, compassion, inner peace and unconditional love. It is also an immaculate tool to aide with out of body experience practices if you are currently working with such advanced levels of spirit work. Even if you are beginner, results can and will eventually be accomplished through dedication and routine use. With commitment and repeated exposure, anyone can experience the transformative effects that this bio-field programming was built to deliver. In combination with PMP 3.0 which essentially offers the same results, the experience has been nothing short of superlative. What results could it produce for you? You will never know unless you try it right? Tick-tock. Your spiritual awakening awaits you in the world of bio- field technology, sonic resonance, harmonic frequency and metaphysics. Infinity Lambda Brainwave Meditation was designed by an American named Douglas Prater. He has a degree in music and sound recording technology from Texas State University. He is an author, a musician, meditator and a fitness enthusiast. Deep meditation practice and spiritual evolution are key factors of his life. I have great appreciation for the work that Douglas Prater has done creating this outstanding bio-field technology system. It delivers results on all levels promised. As noted, Lambda brainwaves are associated with the ultraviolet causal chakra above your crown chakra, outside of your body. Epsilon brainwaves are associated with a burgundy base aspects causal chakra that is actually located off the human body, positioned right below the feet. These two brainwave states when regularly empowered and activated, are a prime component of achieving mental transcendence. Consciousness must be expanded to this capacity in order to achieve successful alignment with any intelligent spirit force residing in the collective mind. The connection must first be made to inner self through the heart chakra. For myself, the ascensional journey into the higher states of consciousness takes the path of *inward* followed by *upward*. This information is a critical key.

The universal consciousness is omnipresent within the vector field. In other words, this all knowing spiritual intelligence resides all throughout the fabric of the higher vibrational, energetic densities. We may refer to this space as the akashic field, the collective mind, the vector field or the quantum field. All of these descriptions are quite acceptable. This spiritual plane may also indeed be the realm of the Absolute itself. The universal consciousness is everywhere all around us, above us and below us. The collective consciousness is a vast bottomless ocean of endless, infinite wisdom and knowledge. Unlimited amounts of information and volumes of precious data are stored here. When perceived from the proper vantage point, it's literally like a metaphysical super computer. The memories of knights, kings, dynasties, heroes, masters, Saints and legends are part of its programming. Endless soul forces and the conscious experiences of multitudes of generations of humanity and ascended angelic beings all reside here. Their data has all been encoded into the system of the collective mind. All information regarding the scientific processes of Nature and origin of creation models are also programmed into this system. By tapping into this realm or tuning our chakra and brainwave frequencies into this channel we can potentially access, process and record unlimited volumes of information in relatively limited periods of time. In my case this phenomenon is more or less experienced as a massive download of information that comes in at speed of lightning. The data must be captured and recorded through handwriting. The information cones in all at once within my consciousness field upon my stream of thoughts and it only stops when the message is complete. A massive amount of information can come in one tightly packed thought. An entire book can be written on this way.

Both of these magnificent consciousness expanding bio-field technology systems combined as daily alternating sessions were key factors in my spiritual evolution. My observable, tangible, intellectual transformation and the development of a familiar, personal relationship with my Inner Self has been beneficial to my mind, body and soul. The Inner Self is always connected to the higher sprit forces of nature that exist all around us. This tie is never severed during each incarnation. Most people will live entire lifetimes never seeking out this aspect of themselves. In most cases this is only because they are unaware of this divine presence with them. As such, the initial key requirement is to acknowledge its presence. Next, it is imperative to establish a solid relationship with this higher aspect of angelic spirit that resides within you. In this way you can access spiritual realms that vibrate at the higher frequencies, which your inner self has the natural ability to automatically attune to. Its that simple to comprehend. Both of these outstanding mind expanding brainwave entrainment systems can facilitate this process. They can be found at www.iawaketechnologies.com .

What does our creator expect of us? As members of the human race it is our obligation to acquire as much knowledge as possible in each incarnation. As rational, biological beings we are expected to follow a code of conduct like the 10 commandments, Universal Law, the Golden Rule and so on. It is our obligation to display proper ethics. The strong are to protect the weak. We are to be respectful care takers of a planet that is on loan to us. The kingdom of mankind is to respect their fellow mammal and animal kingdoms of the land and sea. These are the standards set forth by God-Mind, our Source and Creator who answers to so many names. Our main vocation as humans is to continuously develop our intellect, compassion and spiritual awareness. In addition, at some point in our lifetime we are naturally destined to experience the great, expansive, divine force of unconditional love. These are the reasons why we are truly here. Consequently by following these basic, standard guide lines each of us makes a personal contribution towards becoming the best version of ourselves that we could possibly ever hope to become. In this state of being we have something to offer the world around us. We become an asset in the ongoing, continuous evolution of the human race. Each of us has a destined service to render to humanity and to our planet. It is our obligation to do this selfless service with gratitude for the opportunity.

When in doubt simply ask yourself, what would Jesus do? After all, he set an immaculate standard on human conduct in a physical, spiritual, mental and intellectual manner like no other who had ever come before him. That is a part of the reason why the whole world knows that he was here. He was here for all of us. Everyone of us despite age, race, religion or creed was his target for spiritual reformation and enlightenment. He had a message. His information was good. But his true life and true history are literally shroud in mystery. So if you want to know what Jesus would literally do in any given situation you must know who he really was as a person. His incarnation was divine, he became a *Sun of God*. He achieved a state of transcendent consciousness that elevated him to the status of a Sun God, a solar deity that left his mark on all of us. Sadly, his presence here was slightly underappreciated as we all know. How ashamed embarrassed should we be as a race for the persecution and destruction of such a genius, loving saintly spirit? Let us know and remember him for who he really was, why he came here and what he scientifically, spiritually and mystically accomplished for the human race. Blessings to our Ascended Master. He who came here through the love and radiant light of God our Creator and Source, to heal and educate us. We are expected to follow the code of conduct that this ancient soul and angelic being has laid out for us. We are to naturally desire to live in an enlightened state of Christ Consciousness. The new age is an age of reason, unconditional love, radiant light and mind blowing technologies. Amen.

When any member of the human race experiences a spiritual awakening the energy that is generated and emitted outward, reaches others in similar states of consciousness. This effect occurs and affects our state of mind even though most people are often unaware of it. The inherent oneness that connects us all allows for the energy of each individual awakening to contribute to the awakening process of other members of humanity. This energy transfer reaches anyone who is actively attempting to reach the same levels of consciousness that someone else has already attained. Why is this so? Because everything is connected. This point cannot be stressed enough. As such, making the free will choice to become the absolute best version, most awakened and spiritually transcendent person you can be, is the greatest act of service that you can render to humanity. Your expanded essence goes out like a signal to everyone all at once. When any one of us attempts to make healthy choices, be ethical, kind to others or simply express love as an aspect of creation, the higher consciousness levels that some of us have attained empowers these forces. Fundamentally, this promotes a more spiritual state of existence for the whole. On a metaphysical level, this is how Nature works because all is one.

What we do to the one, we do to the all. It is a simple concept to grasp. Are you beginning to understand why aligning our energy to natural law is so vital, critical and significant? As biological lifeforms we are connected to our creator through infinite consciousness and vast, universal forces of unconditional love. This is the ultimate essence of the mind of our Source. Never underestimate the power of the human mind. When properly trained, properly directed and in perfect alignment with universal law one naturally begins to resonate with the spirit forces found all throughout Nature. When we are in harmony with nature, and in harmony with others around us we begin to care about everything on a much deeper level. We begin to care about out earthly abode and the condition of it. We begin to care about doing our part in making the world a better place for all of us to live. We begin to care about building communities centered on unity, self sufficiency and oneness. We begin to wonder if we are not alone as a species in our universe. Combining spirituality and science sends innocent, spiritually inquisitive pilgrims off to research the glorious metaphysical scientific processes of nature which quintessentially reveal to us, the creative abilities of our Source.

In a new age of light seeing cosmology, numerology, paleontology, astrology, astronomy and plasma physics in the classroom could create a whole new world of explorers. Sharp creative minds who comprehend the glory of our spiritual, material existence and the purpose of it can pave the road to a brighter future. My energy healing program and the tools I used was a conglomeration of diligent research and a dedicated testing of the science over a prolonged period of time. Utilization of frequency, sympathetic resonance, knowledge of the solar spectrum and bio-field technology are at the core of my program. I have an incredibly long history of experience with spiritual metaphysical practices, meditation, creative visualization techniques and so on. So it is imperative to note that I did not start this energy healing program as a beginner by any means. Additionally, I have been working with brainwave entrainment for over 3 years now consistently. I have experienced many positive, effective results. It is true that this program will produce some form of results for everyone. That's a fact. If you seek spiritual ascension and a means to arrive there, bio-filed technology is the one stop, all encompassing real deal. It will help you arrive at your destination. However, the speed and rate of progress at which you mentally, emotionally and spiritually develop will be determined by any prior meditation experience that you already have. Success is also driven by our willingness to shed our vices as we lighten the energy load through the PMP 3.0 Full Spectrum patented releasing techniques.

After 1 year of daily alternation of these two biofield

technology meditation systems, my nervous system began to vibrate in a more harmonious frequency. My mental awareness expanded to incredibly wide range, massive perspectives and colossal macrocosmic levels of thinking and perceiving. I consciously began to observe the really big picture in regard to various intellectual paradigms. As you can see, the first book of this series contains thoughts, contemplations and metaphysical explanations of great macrocosmic origin theories such as the cosmic egg and the holographic universe. In regard to our material existence on the physical plane, it doesn't get any bigger than that. In addition, utilizing these two systems literally cleared my mind of any fanciful, illusionary thoughts, unrealistic beliefs and mental smog in such way that it was simply gone and forgotten. All that remains once unnecessary mental clutter is unburdened is a nice clear mind to work with. Yet, the effects that I am achieving must still be accredited to all tools that I used in my healing process. Each one of these energetic modalities did their part to heal my mind, body and soul in their own specific way. Initially, discovering the proper frequencies to attune my chakra energy centers to accurate solar spectrum sound frequencies was crucial. My energy centers were all desperately in need of recalibration. Perhaps this initial step facilitated in ramping up the rate of speed at which the bio-field technology could take effect. Starting out in a primed energetic state is logically an asset. So I wish to pay my respects to all energy healing tools that continue to do their part in aiding me to ground, center and activate specific brainwave states. In combination, all of it has been of assistance in transcending my consciousness to commune with inner self and greater spirit. A personal connection to the unconditional love of the Absolute, our Source, is the greatest gift that biological existence can offer. Every physical journey is a quest for this divine immaculate light.

Choosing to develop a spiritual bond with Nature and

our metaphysical mother earth, Gaia, deepens and empowers our connection to our heart chakra energy. This in turn, encourages us to be peaceful, compassionate and observant. In this state of being, we naturally begin to witness our external experience from within as a silent observer. Transcending or elevating our consciousness promotes the development of a profound sense of self awareness, ultimate destiny and life purpose. Our ability to achieve deep states of enlightened consciousness are stepping stones to the spiritual evolution our species. The establishment of united, peaceful societies is ultimately a dream that every one of us shares and hopes for. The vote on that is unanimous. Generating an expanded mental awareness and a deep, spiritual sense of presence in the now, contributes positive harmonic energy to our race as a whole. To believe otherwise would defy Natural Law. For as we know, all is *One* unified life force. Though I have mentioned that point numerous times already, it can never be stated enough.

Each and every one of us can participate in raising our own vibratory frequency, developing compassion and expressing unconditional love on the physical plane. When we choose to do so, our sympathetic resonance is sent out to others as a positive harmonic. In other words, biological lifeforms that temporarily occupy the same space we do can feel our good vibes. When we express joy, happiness, unified feelings of oneness and spiritual energy this force can indeed be felt, witnessed and experienced by those around us. Our positive vibrations reach them through resonant bio-filed energy exchange and others can absorb our positive energy if needed. Fundamentally, this possibility always readily exists because that is how nature works. Likewise, we can mentally and emotionally choose to block negative energy that emanates and imparts from others by using various, simple energy protection methods. Whenever you are exposed to negative energy or negative people, basic shielding techniques are an essential, critical tool. Not allowing negative energy emitted from others to affect you is a simple matter of learning how to consciously block it out. Divine physical connections to nature will always deepen our experience as spiritual beings having a human biological experience. We can feel and sense loving energy both within, without and all around us. In fact, the force of universal love is omnipresent.

Once we develop a common state of higher vibrational energy and internal peace, it is imperative that we intentionally extend this loving energy outwards into the world around us. The more we choose to willingly do this, the easier it becomes. The effect of this cause is the ability to connect to everything in nature at will. This state of existence promotes expressions of universal unconditional love and proper human conduct. It further empowers positive ethics, co-operation, expanded intellectual thinking and proper action. It gives us a natural, automatic desire to extend selfless service to the human race in an effort to play our personal role in making the world a better place for everyone. In combination, each of these energies aligns to Universal Law set forth by our Source as we know it by way of harmonic resonance. To get to a destination filled with light, we must inevitably first walk through our own valley of the shadows of darkness. We must embody the great powerful strength and endurance witnessed within the hearts of the Israelites, led by Moses as they forged their way forward in the treacherous, smoky darkness of the days of old. At this pivotal juncture in linear time, my life experience has more than adequately taught me this fact. Within this darkness, each of us must face our own inner demons, weaknesses, vices, defects and obstacles. Fearlessly, we set to the task of valiantly defeating, overcoming and annihilating these unnecessary obstacles. In order to release and permanently let go of and shed our vices, we must logically deactivate all outdated, useless programming. Finally, we reprogram the system with immaculate new, updated programming. We release and reprogram ourselves with bio-field technology and in effect upgrade to the 3.0 version of ourselves. That is how it's done. How does it feel? In a word, Amazing.

Profound Meditation 3.0 Full Spectrum contains the necessary entrainment technology required to help practitioners release energy that was built up and stored within the nervous system over the course of an entire lifetime. How remarkable is that? Any unwanted energies are drawn out and immersed in a potent energy solvent that helps us dissolve all obstacles to spiritual growth. This systems unique structure was designed to completely cleanse ones mind, body and soul. IAWAKE bio-field technology was built to purge and remove dysfunctional emotions, outdated misaligned beliefs, anger issues, addictions, stresses and compulsions. If you suffer from any other these conditions PMP 3.0 Full Spectrum has the ability to locate and target those energies within you. This program was designed to aid in removing and overcoming obsessions, unhealthy thought processes, bad habits, doubts, fears, judgmental energy or any other limitations to self development. This effect is experienced as a feeling of negative character defects being drawn out, cleansed and lifted away. This energetic release process occurs continuously with ongoing use of the system until a perfect harmonious, balanced state of being is achieved. You will know when you have reached this state. You will literally feel and sense the difference for yourself. The power of PMP 3.0 Full Spectrum and the results it produces, speaks for itself. The bottom line is that all negative energies must ultimately be released if one truly ever seeks to properly transcend consciousness. This is easier said than done especially if our environment is unhealthy, negative or unproductive. These types of circumstances actually give us greater cause to heal internally. No matter what type of surroundings we should find ourselves in, releasing bad energy is by no means impossible. All that is required is deeper commitment to healing and a more focused effort. Devout desire to expand our awareness is a prerequisite to elevating our spiritual vibrational frequency. Dwelling in an overall higher vibratory state of being generates a healthy, balanced sense of harmonic existence.

Operating at top levels of efficiency facilitates peak performance, increased productivity and a sense of global unity generated by our heart chakra. These positive forces are naturally inherent within us but they are only accessible to those willing to work towards self mastery as a state of existence. There are an abundance of tools available to those who are willing to make the effort. Additionally, the spirit forces of Nature naturally aid those who sincerely try. Once we generate a spiritual awakening we inherit a much deeper connection to inner self and the ability to access this divine quintessential aspect of ourselves much more easily as we wish to. Eventually we reach a state of proficiency that allows us the ability to alter our vibrational frequency at will. This is a powerful spiritual sensation to experience. We turn our attention inward towards the internal, subjective, spiritual self, as this aspect of our being is the true essence of our immortal soul which embodies the connection to the one unified consciousness. The one taking all these biological journeys on the physical, material plane is the soul spirit on the inside. The outer, objective ego self is only one of the many fleeting, biological personalities belonging to an earth vessel that houses our consciousness, soul and spirit while we are here on earth. The memory of the entire life experience of our objective personality downloads into the hard drive consciousness of our internal self. Our outer personality can only become the internal self of a new external incarnated personality by becoming an ascended master. Who is the most famous ascended master of all? A dearly beloved Saint we all know of named Jesus. Amen.

We are one with all things around us, yet while trapped in these physical bodies, we are locked into the limited illusionary state of division. This false sense of separation from other things in Nature and other members of the human race, is a result of our so called individuality. By remembering, acknowledging and reconnecting to Universal oneness in our hearts we deepen our relationship, with our inner self, the holy guardian angel within us. It is through this gateway of the heart that we initially gain access to the universal collective mind of the All. This is the path. This is the road. This is the route to the light of God-mind. This is way to the abode and realm of saints, ascended masters and angelic beings. Saviors and Messiahs reside here also. Energy healing tools can greatly assist us in reaching states of transcendent united consciousness. That being said, the real magic of it all can only be achieved and activated by personal will empowered by desire and directed intent. Spiritual transformation and transcendence of consciousness require the humble dethroning of the ego self. This action paves the way to the intelligent acknowledgement of the existence of the angelic soul within. This essence is our ultimate center of being. As this immortal aspect of us is always connected to the cosmic mind, to acknowledge and accept its existence is key. Once we have done this, we must trust and permit this force to guide us wisely. In this way we come to express the heart centered love and light of Christ himself.

The most impressive thing about using PMP 3.0 Full Spectrum and Infinity Lambda Meditation systems is that long after the session is complete the raised energy actively stays within our bio-field. This effect has been intelligently built into the blueprint and design of these systems. The energy of this program saturates the practitioners bio-field, physical, astral bodies, nervous system and auric field. Long after the session is over the energy continues to resonate at a noticeable subtle level. This effect allows us access to this energy as required throughout the day. No other system can offer this profound feature in the same way. As such, we have the ability to utilize this energy field at any point when needed. In this way we can instantly dissolve and consciously release any negative unwanted energy that randomly, naturally arises. There will always be stimuli and stressors to react to in this world. We can't change that. We can only change how we respond to them. These may include such energetic resistances as outside emotional stimuli, random discomforts, unhealthy attachments, mental aversion or negative influences.

We can immediately alter how we respond to the energy of others and how it affects us. We can instantly acknowledge and deflect self defeating thought patterns, anxious energy, tension and mental obstacles. The list goes on and on. Any mental or emotional reaction to the conditions within our environment ultimately indicates lack of balance. Learning to identify, label this energy and correct it on the spot is critically essential. The vital success of spiritual awakenings, advanced states of consciousness and mental transcendence fundamentally depends on our ability to release all energy blocks. This is a critical, mandatory requirement. The necessary action of letting go of hindering blockages that prevent spiritual growth and development is the only way to achieve effective results. Letting go of all the things that we are *not* is the only thing that allows us to discover and embrace who *we really are*. Once we release and shed all unnecessary internal debris only the good within us, filled with the light of our Source remains.

Once our character defects are recognized, confronted, diffused, permanently removed and purified these things can do no further damage. A spiritually clean slate promotes growth, evolution and transformation. Cleansing of negative energy will always raise our biological vibrations to healthier states of being. Higher vibrations increase our physical energy levels and empower our vital life force commonly known as Prana or Chi. Harmonic frequencies that vibrate in harmony with Nature enhance the functions of our chakra energy centers. Exposure to proper brainwave frequencies activates our entire sympathetic nervous system and specifically promotes our spiritual connection to heart centered existence. Everything connects. In regard to energy healing, this critical detail is at the heart of the creed. Increased levels of vital life force provide endless spiritual benefits. Increased energy flows help us generate higher states of awareness and give us the ability to access transcendent states of mind more easily. Negative unwanted energies will always present as blockages until they are acknowledged, addressed and released. All it takes to initiate a progressive spiritual evolution are the proper tools, correct knowledge and the right attitude.

At the end of the day, was buying Profound Meditation 3.0 Full Spectrum and Infinity Lambda Meditation worth my money? Hands down, 100% yes. This revolutionary brainwave technology is superlative by way of construct and design. Hats off to John and Pam Parsons Dupuy and to the entire IAWAKE team. Much love to all of you. Your contribution to humanity is a priceless energetic asset that was desperately required. Fellow members of the human race, I welcome you to the long awaited dawn of the great awakening. For those of you who can already feel it every day, spread the internal joy and love that you house. Welcome to the light of the incoming Aquarian Golden Age of reason. Great transformational forces are already at work in Nature as our new age rapidly crowns the horizon. The New Age community of spiritual practices and traditions is exploding with metaphysical truths, sympathetic frequencies and good information. There's a reason for this intense world wide interest in enlightening knowledge at this time. You see? The spiritual force of the awakening speaks for itself. So yes one thousand times over and two thumbs up for these impeccable bio- field technology systems and the incredible effects that PMP 3.0 and Infinity Lambda generate. Despite this top drawer review, please don't just take my word for it when I tell you that this system works. Instead, put my recommendation to the test and try it out for your own self.

This technology is a key. It has improved my entire overall attitude and observable outer personality. It has aided in increasing my spiritual awareness in ways immeasurable. The benefits of this system are far too numerous to count. I continue to use it on a regular basis and it continues to provide increased effectiveness. Sharper thoughts and mental deduction processes are continuously upgraded. IAWAKE technology doesn't eliminate the many other energy healing tools that I align to and naturally resonate with. But this system does get to wear to the crown and hold the scepter. I grant it the status of harmonic Royalty. My crown jewel. That being said, it is imperative to acknowledge the many different modalities of energy healing, tools, techniques, practices, frequencies and so on that help us heal in different ways. Each of these individual modalities of energy healing help us to achieve different states of spiritual presence and new levels of mental awareness. Some systems are more accurate than others. Some may be very effective for you while others may not produce the results you seek. Some work exactly the way the designers intended it to and some don't work at all.

The frequency A432Hz is omnipresent throughout nature and creation as the cosmic frequency of the Golden Ratio. Our body, like all other things in creation, is comprised of water and crystals. Both water and crystals store information. In effect, every vibration that we hear becomes stored as information inside our bodies. As a result, most of what we listen to throughout the day consists of inharmonious, dissonant and unhealthy harmonics. Inharmonious frequencies generate negative effects energetically, emotionally and spiritually. If you read the science section of my 1ˢᵗ book *Awaken and Ascend* that reviews and expands on the teachings of Dr. Gerald Pollack regarding the 4ᵗʰ phase of water, you will learn the science behind the ability of water and crystal to be programmed with and to store information. As you can see, the accuracy of the frequencies that we expose our brainwaves to must be precise and accurate if we wish to achieve successful results in energy healing.

With access to proper energy healing tools, advanced practitioners can take their consciousness to whole new levels. Novices can also become experts and masters of the self with continued exposure to IAWAKE bio-field technology. The feelings of inner peace, generosity, kindness, forgiveness, unconditional love of all things in nature, expanded mental states and spiritual connectedness generated by these systems benefit everyone. Anyone can establish a new sense of reality, a new state of being and an elevated state of consciousness. To express an interconnected sense of unity and oneness with all aspects of biological creation is actually part of the natural, inherent human blueprint. It is the ideal state of being that we ultimately strive for. By connecting to our Inner Self within, we gain the ability to connect to divine forces of spirit outside of ourselves in higher realms. We can accomplish this goal much more easily and effectively and with greater ease by incorporating bio-field technology into our energy healing work and spiritual practices. There are endless benefits and powerful effects associated with the mind blowing scientific bio-filed technology systems that have been produced by the amazing IAWAKE harmonic architects.

In the next and final chapter of this book we examine a simple method that can be utilized to entrain our brainwaves and heart rate at will, using only mind power. Techniques such as this require mental awareness, memory recall, the ability to concentrate and the power to direct our consciousness. Doesn't that sound intriguing? Let's move right along and find out how its done.

CHAPTER 13

Mind Over Matter. Learning How To Alter Our Brainwave States And Heart Rate At Will.

In this final chapter we will be reviewing a simple technique to accomplish instant brainwave entrainment whenever necessary. The following information is an expansion on the research of Dameon Keller. What is presented is a very simple method can be practiced any time, anywhere. This technique is easy to master and will work whether you are alone or surrounded by people. For further study of the material and subject matter presented in this final chapter, please visit energy worker, sound therapist, musi producer, Dameon Kellers official website, published work *Sounds Good, Sounds Great, Sounds Amazing!"* www.dameonkeller.wixsite.com.

The most popular, way to entrain our brainwaves to certain beneficial healing frequencies and specific desired states of consciousness is through the use of binaural beats and pulsed isochronic tones. Listening to pulsed beats and isochronic tones designed to stimulate certain brainwave states will affect and alter our consciousness almost immediately. Meditation practitioners who have less experience working with frequencies or no previous exposure to binaural or isochronic tones tend to exhibit a change in attunement after 6-8 minutes of listening to an audio track or meditation session. After the session is complete, the after effects will often last for approximately the same amount of time as the duration of the session itself. Essentially, if you are exposed to a 1 hour binaural session in the theta brainwave range, combined with 7.83Hz Schumann Resonance the effects of the raised frequency can be felt and experienced for up to 1 hour after the session has ended. So how do we achieve this mental state without physical exposure or immediate access to the brainwave entrainment audio session? The answer to this riddle is all about mind over matter. This skill requires focus and excellent memory recall abilities.

To achieve controlled entrainment without access to the brainwave entrainment audio session, one has to internally produce the effects of the absent physical stimulation tool. We must operate from the inside where our memory of the binaural or isochronic frequency is stored if we wish to regenerate it. When we download any kind of information and develop a comprehensive understanding of whatever we have learned, this information becomes a program on a brain cell. An electrical neuro- transmitter will fire over to the brain cell that carries that program, the more we focus on that memory which has been stored in our mental database. This is one of the many amazing features of our conscious cerebral functions. Utilizing the faculty of memory we are able to recall the pulsed binaural beat or isochronic tone and alter our brainwaves almost instantaneously. Regardless of where you are at the time of need this process can be performed right on the spot. This method requires little more than three minutes to take effect. Ultimately, this simple unique procedure produces the same results that would be generated when we actually listen to the entrainment frequency externally. By using this method we are still hearing the frequency but we are listening to it in our mind, from the inside. In this way, we experience the harmonic entrainment tone internally as opposed to externally. Acquiring this skill serves multiple spiritual and physical healing purposes. Additionally, this technique can be used to contribute to balanced, grounded energy anytime by providing instant access to a powerful healing tool as required.

Learning how to alter our brainwave states at will is all about the power of the mind. This practice requires focused memorization and the ability to recover the memory of the frequency from our memory archives. In other words, our ability to tune into the brain cell that contains the programming that we are looking for. When we begin to think about something that we already know, we instantly activate a neuro-transmitter, an electrical signal in our brain. This signal intelligently goes out in search of the information we are seeking and thinking about. This process occurs naturally, as it is an automatic biological function of our physical body. Thoughts are being transmitted throughout our brain at the speed of light at all times and this function is always occurring despite the fact that we don't even notice it. We also have the ability to tap into this amazing natural force at will as required when it is necessary to recall anything that we desire to remember. Memory recall of this nature is a main function of our instant, automatic brainwave processes. All we have to do to tap into our reservoir of wisdom at will, is pull the memory of the frequency or any other memory of choice forward in our mind.

When practicing this technique of instant brainwave entrainment, it is advantageous to first pick a powerful audio track that meets your personal, spiritual healing requirements. Next, listen to the session as frequently as possible. Do this preferably every day, but it is highly recommended that you do not exceed 1 hour of entrainment frequencies daily. This precaution is set in place so as to avoid over stimulation of our brain, nervous system and energetic overload of our other biological faculties. This limit is a standard that applies to most of the frequency entrainment audio systems and programs that are available to us. Listen to your chosen track enough so that you have thoroughly familiarized yourself with it. Do this daily until, you have memorized the entrainment frequency session to such a degree that you can literally play it in your head.

We use this simple method unknowingly on a subconscious level when a song that we love gets stuck in our head. When we have heard a song a great number of times, we do memorize it even if we are unaware of that. When we recall the song from memory we can almost hear the song playing in our head from beginning to end along with the lyrics and the rhythm of the beat. Ultimately we can consciously employ this exact same method in regard to brainwave entrainment and heart rate control. All we are actually doing, is recalling a memory at will. Step 1 is to choose one track to meditate with and expose yourself to it repeatedly until you know it like the back of your hand. The more experience you gain utilizing brainwave entrainment systems the easier it becomes to recall these frequencies from memory where they are infinitely filed and stored in our vast memory data bank. The more you listen to an audio entrainment session the more accustomed you become to it. Once you have it memorized, like a song or commercial that gets stuck in your head, you can play it your mind whenever you wish to. Using this technique you can activate certain brainwave states to reach specific desired peaks at will. Initially it is critically significant to memorize the binaural beats or isochronic tones and frequencies that comprise the session. Next we must willingly commit to recalling them at some point in the future. If you have memorized it properly, you have the ability to recall this tone even if you haven't pulled it forward in your mind for months. When you suddenly remember a song that you haven't heard in years, the same faculty of memory recall is in operation.

The act of clearing ones mind of unnecessary clutter and useless information facilitates our ability to focus, concentrate and input good programming. It is of tremendous assistance to simply turn off, eliminate or at the very least, dramatically limit any exposure to aggravating, dissonant, unhealthy frequency ranges. Frequencies such as these have a tendency to purposely stimulate your brain in a negative manner. To avoid these unwanted, harmful, subliminal stimulants turn off or drastically reduce your exposure to television. Stop watching the news. Most of the time it's bad news anyway. Every once in awhile Google headline line world news just to stay on top of global events as ignorance is actually not bliss at all. It's more like obliviousness and that mindset accomplishes nothing in the long run. By reducing your exposure to television, you limit your exposure to musical, lyrical harmonics that were actually very cleverly designed by professional advertisers to entrain your brain whenever you see and hear it. Advertisements are often comprised of frequencies, tones, words and harmonics that produce negative aggravating intervals and rhythms. We are constantly exposed to external stimuli such as this through mediums such as television, radio broadcasting and most mainstream music. Most of the mainstream music that we hear is more often than not, recorded in a pitch like A440Hz that does not resonate with the solar spectrum frequencies that correlate to our chakra energy centers like A432Hz naturally does. Industry standard is to tune to the commonly known A440Hz pitch. Therein lies the inharmonious issue with a lot of music on todays market. Enough said.

If it is possible to eliminate exposure to harmful, distracting external stimuli, the mind will naturally begin to attune to higher states of awareness. Your brainwaves begin to align to healthy balanced states of sympathetic resonance with everything in Nature. The sharpness of our brain functions and performance automatically improves and begins to operate at higher efficiency levels. Removing or eliminating harmful, negative, distracting external sounds, dissonant harmonics, tones and rhythms will clear the noise from the mind. This action promotes calmness, stillness, centered inner peace, compassion and unconditional love. Essentially the ideal solution to removing distracting mental clutter is to turn of the stimulus completely. I turned off the television over 3 years ago and now my state of mind is entirely different than it previously was. My thought processes are abundantly clearer and much more focused on scientific academic study, alchemy, metaphysics and energy healing. Turning off the noise eliminates the problem of mental decay. In short, an abundance of television will produce the opposite effect. Excessive exposure numbs the mind to sleep. Sadly, for a lot of people, just to even imagine shutting the television set off permanently could result in emotional crisis or cause a partial nervous breakdown.

If the utter thought of turning off the television invokes instant separation anxiety within your biological earth vessel, take a deep breath. There is a solution. If you cannot fathom life without television or common industry standard equal tempered music in an inharmonious A440Hz pitch and other radio noise, at the very least, limit or reduce your exposure by 50% and buy a good book or two that elevates and stimulates your mind. The book can be on any subject of interest to you as the importance of the act of reading itself exercises our alpha mind in a productive healthy manner. Reading, studying and learning actually does increase brain function efficiency. After an extended period away from this annoying mental distraction I am of the firm belief that continuous exposure to television has the ability to turn your mind into a vegetable. You only truly notice this madness after you've turned it off completely for any great length of time and reflect back on this interesting matter in hindsight. In any case it is of significant value to simply be aware of the harmful effects that these signals and stimulants have on our brainwaves and subconscious minds. It is imperative to note that there exists an immense amount of really excellent programming which can be accessed through online sources. You can find endless educational documentaries online relating to every field of academic study imaginable. There are endless books which can find online that come from every corner of the world. There is so much wisdom to take in all around us. Why waste your brain staring at box , watching commercials, soap operas and sitcoms that are entraining your brainwaves and directing your thoughts in a negative manner unbeknownst to you?

Being consciously aware of the negative impact that these forms of mental stimuli have on our mind, body and spirit is step one in reducing the harmful effects. Once we become aware of something it cannot continue to influence our subconscious minds in the same manner in which it previously did. Not unless we knowingly and willingly permit it. Clearing the clutter only leads to higher states of mind as well as improved states of physical, emotional and mental health. Ultimately, this is the stuff that logical, calm grounded energy coupled with open receptive consciousness is truly made of. To expand mental awareness it is necessary that we familiarize ourselves meditation practices, healing frequencies and different types of brainwave entrainment techniques. We can utilize binaurals, isochronics or bio-field technology to raise our harmonic resonance and conciousness levels. Listening to properly tuned, just tempered music is very melodic and healthy to the human ears. The more we expose ourselves to proper sympathetic frequencies especially alpha frequencies associated with ocean waves and water in motion, the more we become aligned to Nature. Connection to nature automatically encourages further alignment with our Inner Self. This facilitates attunement to the collective as well as an establishment of a much more energetically balanced state of existence. When properly balanced, mind, body and soul spirit automatically become healthier, spiritually uplifted and our sense of awareness gravitates towards the positive.

All of this generates a very excellent state of being. It's the exact optimum set of conditions what we all truly strive for. Improved brain functions and cognitive thought abilities automatically improve after only a few weeks of exposure to brainwave entrainment techniques. Regular use of binaurals, isochronics and bio-field technology that properly resonates with our biological and sympathetic frequencies has infinite benefits. Being properly harmonically attuned leads to better, more relaxed sleep, experiencing less illness, decreased fatigue, less physical pain and can even aide with migraine issues. Emotional imbalance can also be rectified.

In order to remember something that you need to recall simply clear your mind, take 3 deep rhythmic breaths and go to the place in your mind where the information is encoded by simply thinking about it. The act of thinking about it, activates a brainwave that causes a neurotransmitter to fire off in rapid search for the programming you seek. Which it will find if the information has been memorized and committed to memory. That's a very basic cerebral function that our brain is actively participating in at all times. Commanding or instructing our brain to recall something isn't much to ask of it. This memory function is an active process at all times regardless of whether or not we try to cause it. After only 3 minutes of direct concentration on any targeted frequency, our brainwaves, heart rate and human bio-field respond to the mental stimulation. This automatically activates the desired brainwave state in the same way it would by actually listening to the brainwave entrainment frequency track. After listening to the same meditation session for days, try experimenting with this method to test your ability to achieve rapid results. As long as the audio entrainment track has been perfectly memorized to the best of your maximum ability you're good to go. This form of induced activation requires willing intent combined with a sharp memory. As long as we choose to remain focused on the desired brainwave state that we generate, we will continue to feel the effects of this mentally induced entrainment. In other words, this condition will persist for as long as we choose to sustain our focus on the harmonic tones of the frequency that we are internally playing in our mind. Doesn't that sound easy? A complete description of the simple method involved in achieving this form of induced mental entrainment is provided in the pages ahead.

At the heart of energy healing, it is imperative that we become accustomed to utilizing frequencies that sympathetically align our brainwaves to our natural healthiest states. We can only do this through repeated exposure. When we choose to participate in this type of healing, noticeable positive changes in our overall sense of being are rapidly produced. Spiritual, mental and emotional transformations initiate under conditions such as these. A body, mind and spirit properly attuned to Nature naturally becomes more balanced and energized with elevated levels of consciousness. It has even been reported by some that brainwave entrainment has a tendency to decrease pain in their body. Sleep, focus and concentration have been reported by many practitioners to improve dramatically. Regular practitioners report increases in energy levels. Furthermore, the effects caused by negative external stimuli like inharmonious music, television ads and other dissonant sounds that we are exposed to throughout the day are often greatly decreased. Raising our harmonic frequency contributes to improved health and wellness all around. In higher states of vibration we empower a more compassionate, spiritually expanded, enlightened and consciously elevated state of being. This contributes to less disruptions in our circadian rhythms and increases the productivity rates of our many other critical, biological processes. As you can see there really are endless health benefits to utilizing brainwave entrainment sessions, harmonic frequencies and bio-filed technology on a regular basis.

If you have any pre-existing medical conditions that may prevent you from being exposed to certain types of frequencies, it is imperative that you consult your doctor prior to engaging in the use of brainwave entrainment tools such as binaural beats, pulsed isochronic tones or bio-field technology. Consequently, certain limitations have been placed on this form of energy healing for people with very specific illnesses and pre-existing medical conditions. Consulting your doctor prior to use is recommended in any case where doubt predominates. Safety first of course. That being said, in most cases where limiting health conditions are neither an issue nor a concern, brainwave entrainment methods are easy to use, highly beneficial to our health and perfectly safe for the average majority of people. For the general healthy majority, this form of treatment is high benefit, low risk. Furthermore, the method of memory recall being described herein has zero risks, zero side effects and zero adverse health effects. Consequently if you listen to a certain frequency every night to go to sleep such as a delta entrainment session, it's logical advice to try not to pull that particular frequency forward on your mind while you are operating a vehicle. Nor when you are piloting a plane, navigating the ocean floor in a submarine or operating any type of heavy, industrial Caterpillar machinery. This good advice applies to all.

Once a healthy entrainment frequency gets committed to memory, you are ready to access this powerful energy healing tool whenever it is required. We can now activate the brainwave state that these frequencies induce anywhere or anytime we choose to. It's far better to have a frequency that activates healthy brainwave states stuck in your head than an annoying dissonant pop song filled with subliminal messages. Television programs and commercials are filled with these types of inharmonious tones and pitches. Subliminal messages can influence us in a negative manner by encouraging compulsive mindsets. For example, many commercials are loaded with influences that drive the body to crave unnecessary snacks and unhealthy foods like fried foods, sugar, coffee and alcohol. As a result specific advertisements influence people to go to the kitchen repeatedly for all sorts of unnecessary treats. Even on a full stomach. This lack of moderation is a provoked influence. There is nothing good at all about conditions such as this in regard to our overall state of health and sense of well being.

It is critical that we take proper care of both our minds and bodies if we seek to live in an uplifted state of expanded, compassionate awareness and higher, spiritual vibrations. Daily lifestyle changes that promote excellent health as well as a well balanced mind, body and spirit are key. All healthy routines require some form or meditation as a vehicle to quiet the objective mind. The addition of sympathetic frequencies that resonate with our biological body, spiritual astral body and brainwave frequencies propel us into higher states of awareness and vibration. Regular exposure to brainwave entrainment tools facilitates a better control over our mental and emotional faculties. Removing sugar, salt, coffee and alcohol from our diet ultimately improves our physical health while aiding in removal of physical energy blockages from our biological earth vessels. Eating an abundance of fresh, raw foods, fruits, vegetables and healthy forms of protein are essential. Drinking plenty of shungite water is the ultimate aide for boosting our energy, our immune system and energizing our physical, emotional and mental health. All of these suggestions will serve to empower our ability to gain control over our own physical, mental and emotional faculties. Getting plenty of fresh air, proper rest, sleep and as much sunlight as possible promotes optimum healthy connectedness to Nature and radiant states of being. Practicing routine brainwave entrainment meditation helps us to establish inner peace and internal spiritual power over the external, physical ego self. This results in a strengthened ability to instantly access specific brainwave states along with their corresponding heart rate at any time.

It is a proven fact that regular exposure to healthy forms of brainwave entrainment will raise our consciousness levels. Ultimately, that is exactly what they were designed to do. When coupled with bio-field technology the benefits and advantages abound. In combination, these energy healing tools have the ability to generate a profound spiritual awakening. They greatly expand our awareness, elevate our vibratory resonance and contribute to metaphysical and alchemical transformation. To alter your brainwave states and heart rate at will, utilize the simple formula as instructed in the following description.

Instant Mind Over Matter Brainwave Entrainment Method

Seat yourself upright in a comfortable position with your spine as straight as possible. Lay down only if your intention is to fall asleep. Relax, take three deep breaths in through the nose and out through the mouth. Clear your mind by emptying it of thoughts as you breathe. This creates a quiet space in your mind to work with. Next, try to recall the brainwave frequency that you are attempting to target. Simply by doing this, you are consciously instructing the neurotransmitters in your brain to search for the cell that contains this programming. We conduct this task by holding the priority entrainment session tones in our mind as a thought and continuing to focus upon it. Remember it is important that you choose an entrainment frequency that you have already memorized and have willfully committed to memory. Once you internally locate and lock onto to the entrainment session in your mind, you will need to keep your focus directly on this frequency as you listen to it play inside your head. As you continue to hold this frequency in your mind, your brainwaves will automatically begin to attune to the corresponding brainwave range within 5 minutes max. Usually this process requires only 1-3 minutes to generate the induced entrainment depending on the practitioners skill level. Suppose we are seeking to stimulate our mind in order to promote advanced levels of thought processing in order to help solve a problem at hand. For this, we will need to entrain our brainwaves to a frequency in the Gamma 1 brainwave range between 30-40Hz or the Gamma 2 brainwave range between 40-100Hz. Our crown chakra corresponds to the Gamma 2 brainwave states while our brow or 3rd eye chakra correlates to Gamma 1 brainwaves. Visualizing the color of the energy center along with maintaining mental focus on the corresponding frequency can also be highly beneficial in reaching a proper state of attunement. In this case you would picture the colors violet and indigo. That being said, while visualization of the color is a helpful aide during this practice, it is not an essential requirement for successful utilization of this induced entrainment method.

The critical key to success with this technique is focus and concentration on the memorized tone. Once the memory of the entrainment frequency is well within mental focus, hear and absorb the frequency just as you would if you were externally listening to it. Listen to it from within. While playing and hearing the entrainment frequency in your mind, attempt to align the rhythm of your breathing with the rhythm of the harmonic frequency that you are playing and hearing in your head. Once you are successful at matching the rate and the speed of your breath to the rhythm of the tone playing in your head, a new effect occurs. Our breathing, now in a state of rhythmic attunement with the entrainment frequency we are internally hearing, quite naturally causes our heart rate to automatically sync itself to the pace of our breathing. The pace of our breathing aligned with the tempo of the frequency through focus and concentration automatically causes our heart to resonate in union. This quintessential effect naturally occurs because at the heart of it all, everything in creation is intelligently connected. Ultimately, we are willingly, consciously syncing our heart rate to the frequency and brainwave state that we have chosen to mentally manifest through focus. We are literally aligning our energy, breathing, brainwaves and heart rate to a harmonic that we are playing in our head. In effect by hearing it internally we are using our intent, conscious awareness and mental abilities to alter our heart rate and brainwaves at will. This method is literally so easy that this whole chapter could have easily been summed up in 3 words, *Just Memorize It.*

Our heart rate automatically aligns to the rhythm of our focused breathing by way of construct. As long as our focused breathing is aligned to the rhythm of the frequency that is internally being pulled up in our mind and experienced, our heart rate will remain in sync with our directed breathing. In this way we sustain the specific state of entrainment for as long as we maintain our focus and concentration on the matter at hand. Once we achieve success with this simple induced entrainment technique we become more skilled at utilizing it and eventually it becomes like second nature to us. We can then easily access this tremendous force as needed. By way of nature and intelligent design our heart rate, brainwaves and breathing were built to automatically attune in perfect harmony with one another. So there is really no special trick to this method. What we are fully inducing is an otherwise natural state. This process requires no further talent than the ability to memorize a frequency, remember it, focus, concentrate on it and breathe. All that is required to accomplish this task is an initial willing commitment of the frequency to memory and the ability to recall it as needed. Obviously this skill demands repeated ongoing exposure to the brainwave entrainment audio session until its completely stuck in your head. That is when you know that your program has been fully downloaded to a brain cell. For this, we simply need to be capable of concentrating.

Technically, we are picking up subliminal forms of entrainment from various external sources at all times. Additionally, we are subconsciously committing this information to memory unknowingly, unbeknownst to us.
Bearing this in mind one can easily see how using this technique mindfully and consciously with guided intelligent intent and positive directed will, can achieve powerful results. Not only does this technique produce the exact same effect as hearing an external entrainment frequency, it also contributes largely to increasing personal control over your own mental and emotional faculties and brainwave states. This in turn affords us the opportunity to keep our resonant bio-field vibrating at the healthiest frequencies as altering our brainwaves alters our bio-field energy as well. By now, you likely know exactly why this is the case. If you're thinking its because everything is connected, you are exactly correct. This unique method of simple self induced brainwave entrainment and heart rate synchronization can be accessed at will. This technique of memory recall is always readily available and is conveniently free of charge. It is highly effective. It rapidly takes effect and has zero risky side effects.

In my first book *Awaken and Ascend* I recommended various brainwave binaural sessions and harmonic frequencies that I personally use and favor. In this current book I revealed and reviewed my favorite bio-field technology systems brilliantly produced by IAWAKE. While any of these personal recommendations are excellent choices, your choice of frequencies must correspond to your own needs physically and spiritually. However, it is imperative that the frequencies that you choose, attune perfectly and precisely to our commonly known brainwave frequency ranges. Choose entrainment tracks and meditation systems that you feel comfortable with while bearing in mind the tremendous potential, that is packed into advanced bio-field technology. A basic understanding of how our pre-set automatic brainwave range of frequencies correlates to our chakra energy center frequencies is highly beneficial. Our brainwaves correlate to our chakras, sympathetic nervous system and our biological glandular system respectively. If we are to go to great lengths making tremendous efforts to memorize brainwave entrainment frequencies for instant memory recall, we must begin with the right harmonics. It is of critical significant relevance that any entrainment frequencies utilized, sympathetically attune to our precise range of brainwave frequencies. If they do not, no healthy noticeable result will manifest at all. It would be impossible. It would defy the laws of neuroscience if some random frequency outside the range of 30-40Hz or 40-100 Hz were able to attune our brain to a Gamma 1 or 2 brainwave state. There is only one frequency range for Gamma 1 just as there is only one preset frequency range for every one of our brainwave states. That factor must always be considered when choosing the proper binaurals, isochronics or bio-field technology entrainment tools to meet your own specific healing needs. Simply bear in mind that there is only *One* proper harmonic range of frequencies that is capable of stimulating and activating each of our various, specific brainwave states. As such, choosing the right frequencies obviously requires knowledge and comprehension of what our pre-set scale of human brainwaves is. The frequency ranges that generate each specific brainwave all fall perfectly into a flowing, categorized orderly system. Exactly like a rainbow is formed in nature from the colors of the solar spectrum of light, in perfect harmonious order. From a scientific, alchemical and metaphysical perspective, both of these magnificent forces are intricately intertwined.

Each brainwave generates a different type of physical and emotional stimulation and activates different forms of thought processing. Developing a comprehension of the specific brainwave ranges, their main key functions and the different energies they invoke is a vital intellectual asset. That being said, modern neuroscience divides the commonly known brainwave states into 10-12 separate frequency ranges. Each brainwave frequency range produces different psychological effects, states of consciousness and levels of mental awareness along with various other effects and stimulations. In my 1ˢᵗ book, *Awaken and Ascend* I listed our 12 known brainwave states along with their proper corresponding frequency ranges. In this book, this information can be found in the solar spectrum correlation chakra charts in chapter 2. Each brainwave state produces its own unique energetic effects, qualities and characteristics. Furthermore, each specific brainwave correlates to a music note, a gland in the body, a certain color and a chakra energy center. Understanding the energy produced by each brainwave facilitates intelligent selection and proper utilization of precise entrainment frequencies. Only accurate harmonics will generate profound, amazing effects. Choosing these frequencies wisely is key to achieving incredible results in regard to activating specific brainwave states at will. Our brainwaves are always in action on both conscious and subconscious levels at all times, whether we are aware of it or not, asleep or awake. Learning to direct and control our brainwaves at will is an amazing skill to master.

In conclusion I would like to state, that my ultimate goal in writing this book was to uncover the many quintessential connections between Natural Law, the great macrocosmic scientific processes of Nature and our spiritual, biological existence. The Universal Laws of Nature endlessly reflect the constantly changing and developing biological manifestations from the greatest levels of creation to the most microscopic life forms in existence. As a result, the original, primordial cause of all manifested things must absolutely define and reflect the things that we witness and observe in Nature all around us. In regard to the known accepted origin of existence model offered to us by modern science, correlations between the big bang and Natural Law as we know it, simply fail to present themselves. That's because no observational correlations to that singularity exist in our earthly realm. No point of reference to this explosive occurrence can be seen, witnessed or visibly observed repeating itself anywhere in Nature in our physical, biological, microcosmic world. On the same token, the connections between the Seed of Life, the electric universe, the Cosmic Egg and the holographic universe all correspond quite perfectly to every Universal Law and all principles of creation. From the dawn of Genesis to this very moment in linear time everything that we experience as reality is a reflection of another process occurring at a cosmic universal level. This is the governance of the Law of Correspondence in action. Furthermore, correlations between Natural Law and the Flower of Life also abound all throughout our physical, biological world of matter. There are numerous, observable, examples of this truth that we can witness, document, study, research, experiment upon in laboratories and record. On a basic primordial level, all of creation, everything on the physical plane is composed of energy. At the heart of it all this energy is fundamentally comprised of light. In metaphysics this energy was directed outward from the nucleus of the cosmic egg at the dawn of creation. In Eastern Philosophy the source of this light was the vesica pisces formed within the divine architecture of the geometric dyad. The energy then moved through various states of vibration and form which eventually became crystallized into hexagonal matter. This energy is experienced by humanity as an illusionary state of existence in a so called solid world. We call this our reality. Most of us believe that our world is solid and that it is very much quite real indeed. Yet, all things on the physical plane are comprised of hollow particles, atoms, protons, neutrons, molecules and electrons. It's a known fact that as seen under a microscope, atoms are the most hollow materials in existence.

Ultimately the omnipresent energy of our Creator as we experience it on the physical, biological plane is comprised of a chromatic solar spectrum of light in varying states of vibration. These various vibrations cause the light to become metaphysically transduced to frequency. That is the effect of the cause of the Absolute, our Source, who uttered the holy harmonic, and the word was, *Let There Be Light.* We live in a world where everything we see, very much appears to be here, but also it is technically not here. Indeed this is a magnificently profound concept to grasp. For some its paradoxal to even consider. Scientists have been aware of this well known reality for years. This strange concept is scientifically relevant to the electric universe and the holographic universe models on every level. In truth, what we experience as reality and that which we think to be solid matter is not very real at all. In fact it is all an illusion of energy in motion upon which many forces of Nature are acting in order to make all things appear and feel as solid matter. Yet, it is imperative to note that all matter is fundamentally comprised of the platonic solids which are in reality, geometric energy forms comprised of light. Science never ceases to fascinate me beyond the grandest scope of my expansive imagination. The hollow nature of our material biological world is profound scientific, deeply spiritual subject matter upon which to contemplate, ponder and meditate.

In this 3 book series, *As Above, So Below* , so far we have discovered how our Source generates, governs and manifests everything we see in creation through the Scientific processes of Nature which are bound by Universal Law. Absolutely nothing that exists is exempt from the confines or boundaries of Natural Law, including human conduct. With an intelligent, comprehensive understanding of our electric universe of energy we can learn how to attune our resonant bio-field to the Schumann frequency of our earth. Many knowledgeable energy workers, Reiki healers and Shamans regularly attune their sympathetic frequencies to those of Nature. Through an understanding of harmonic resonance and bio-field energy, we can learn to successfully alter our heart rate and brainwave states at will. Through practice and ongoing commitment to daily entrainment sessions, almost anyone is capable of accomplishing this outstanding feat. By successfully utilizing this technique we begin to become masters of our biological earth vessels and our spiritual consciousness. When we are able to control our brainwaves and heart rate at will, we gain more mental power over the objective ego self and our lower vibrational human aspects. Efficiency such as this is an incredible asset when one considers that technically, humans are mammals, unique members of the animal kingdom with cognitive, rational thought capabilities. These gifts were given to us by our Creator because we hold the divine vocation of guardians of the earth and of the animal kingdom. You see, humanity is supposed to protect these things. This truth is built into the construct of Natural Law. When Nature fails us, we are all responsible for it. When any species of our beloved brethren, the animals of our earth become extinct right before our eyes due to causes of the human race, will God not hold us responsible for this effect??? Absolutely, hands down, 100% yes. This is one of the many divine models that are governed by Natural Law. Reaching transcendent states of awareness that encourage ongoing spiritual transformation and growth, is a personal choice. All members of humanity have been granted the gift of free will along with the ability to manifest our thoughts into our reality. In fact, we do manifest our dreams, thoughts, hopes, fears, feelings and actions into our actual life experience every day. All members of the human race are always doing this at all times. In this way, we are ultimately manifesting our own reality. We embody the internal ability to accomplish this creative work as a reflective quality of our Grand Creator, in whose image we have been made. As such, the forces of manifestation and the immense unconditional love of our Source reside within each and every one of us. In this way, human beings have the ability to manifest and generate the type of healthy, positive state of existence that we wish for. This is the real deal that we call our reality. Acknowledging this simple truth and divine wisdom is a key aspect of making the healthy choices that lead to living the joyous life that each of us ultimately seeks.

So in regard to the mission of this literary work, did I find some plausible correlations between Universal Law and macrocosmic Scientific processes of creation such as galaxy formation? Did I find any quintessential, spiritual correspondences in the science presented by Halton Arp that I was seeking? Did I creatively bridge a few gaps between scientific truth, scientific theories and metaphysical spirituality? From a philosophical, theoretical perspective yes I did. At least in my logical observation. The correspondences simply appear to abound.

Will standard model scientists ever agree with and support Halton Arps findings and conclusions on a public platform? It's not very likely. On the other hand there are several professional scientists who do agree with his theories. Many also admire Halton Arps stamina, thirst for scientific truths and resilient devotion to his causes. Regardless of whose right or wrong in this case, the guidelines of mainstream science will always rule the final say. In the meantime, advanced technology such as the James Webb Space Telescope is out there in the cosmos making amazing new discoveries in real time. Are we breaching the horizon of a new scientific frontier or is it already here? Is our pre-destined spiritual evolution indeed intertwined with the emerging technological, scientific revolution? This may very well be the case. Are these two forces of light not already uniting as one to generate the great ascension? Is the Gap between Science and Spirituality progressively being bridged right now in our modern day and age? In many ways, yes it is. Truely, the world is ready for an ascensional, transformational uplifting. Many *awake* and *awakening* ones are aware of the presence of this force. These ones are active participants in this inevitable outpouring of light upon the human race. Have faith that this change for the better will come to pass. For these things too, have been prophesied. Amen.

Marika Sophis

www.ingramcontent.com/pod-product-compliance
Lightning Source LLC
Chambersburg PA
CBHW072349290526
45794CB00001B/44